Differential Equations

Differential Equations

F. G. Tricomi

Translated by Elizabeth A. McHarg

DOVER PUBLICATIONS, INC.
MINEOLA, NEW YORK

Bibliographical Note

This Dover edition, first published in 2012, is an unabridged republication of the work originally published by Blackie & Sons, Limited, London, in 1961.

Library of Congress Cataloging-in-Publication Data

Tricomi, F. G. (Francesco Giacomo), 1897-1978.
   [Equazioni differenziali. English]
   Differential equations / F. G. Tricomi; translated by Elizabeth A. McHarg.—Dover ed.
      p. cm.
   Originally published: New York : Hafner Pub. Co., 1961.
   Includes bibliographical references and index.
   ISBN-13: 978-0-486-48819-6
   ISBN-10: 0-486-48819-5
      1. Differential equations. I. Title.

QA371.T713 2012
515'.35—dc23

2011049274

Manufactured in the United States by Courier Corporation
48819501
www.doverpublications.com

Preface to the English edition

This English edition corresponds to the *third* Italian edition now in process of being printed.

The main improvements in this edition are the following:

(1) the theorem of de la Vallée Poussin (§ 17) on the minimal distance between two successive zeros of a solution of a second-order linear differential equation is now presented in the improved form of Hartman and Wintner (1955);

(2) a new section (§ 36) on equations with transition points has been added, the method employed being that introduced by the author in a paper published in 1954;

(3) the application of the method of successive approximations to linear equations with a non-Fuchsian singular point (§ 49) has been considerably simplified.

I am greatly indebted to Dr. Elizabeth A. McHarg for her care and diligence in the work of translation.

F. G. Tricomi

Turin, October 1960

Preface to the first Italian edition

A book of this kind can have two distinct and almost irreconcilable aims —either to be a reference book surveying briefly all aspects of the subject and containing an extensive bibliography, or, alternatively, to be a teaching book designed to give the student a clear idea of the problems and methods of the theory of differential equations, which is one of the most important branches of analysis.

This volume has been written with the second of these aims in mind, for there is no lack of good recent reference works. It has grown out of University courses delivered by the author, and it makes no claim to completeness. Only questions which can be treated both with rigour and simplicity are discussed, and the subject matter has been further restricted by excluding those topics which demand mathematical knowledge beyond that of the Honours student.

Limitations of space have compelled me to deal only with *ordinary* differential equations (*partial* differential equations are not considered) and to exclude the so-called elementary methods of integration (separation of variables, integration of linear first-order equations, linear equations with constant coefficients, etc.). The actual contents are clearly detailed in the Index; Chapter I is introductory to the following chapters; Chapter II, Chapters III and IV together, Chapter V (the only part of the book which demands some knowledge of the theory of functions of a complex variable) form sections which may be read independently of each other.

Those readers familiar with the principal mathematical interests of the author may be surprised that this volume contains no mention of operational methods, in particular of integration by means of definite integrals. This however would demand more space than is available here, and there are already available the well-known books by Doetsch (60), Ghizzetti (63) and others, on the use of symbolic methods (i.e. the Laplace transform) for the differential equations of electricity theory, etc.

In presenting the work I have constantly tried to stress the importance in the modern theory of differential equations of reading off directly from the equation the properties of its integrals, in contrast to the earlier aim of integrating the equation explicitly. Difficult cases are hardly ever dealt with in their most general form but in the simplest form possible, so that the fundamental ideas underlying the methods adopted may be most clearly seen.

The reader already expert on this subject will appreciate the usefulness and simplicity of the Prüfer change of variable in establishing the existence theorem for eigenvalues (Chapter III) and deriving the asymptotic representation of integrals of second-order linear equations (Chapter IV); also the treatment of the characteristics of a first-order equation (Chapter II) which under the less restrictive conditions used here appears for the first time in a textbook.

I would further point out that in the " asymptotic integration " of linear equations by the method of Poincaré (Chapter V) I have been able to remove the restriction that the independent variable must tend to infinity *through real values*; this allows me to obtain the classical asymptotic series for Bessel functions by a procedure which can hardly be improved on.

I hope this book will be found useful, particularly by the students for whom it is intended.

F. G. T.

Turin, Autumn 1946

Preface to the second Italian edition

Despite the short period which has elapsed since the first edition appeared (1948), this new edition differs in several aspects from the earlier one, mainly because the material of the book deals with one of the most active branches of analysis and within these last years there have been many important contributions in this field.

Several major improvements have been made in the text and many additions to the Bibliography. I would draw attention especially to these features:

(i) The additions to Chapter II, in which topological methods are applied to the study of relaxation oscillations and related topics of importance in non-linear mechanics.

(ii) The deep yet relatively simple discussion in Chapter IV of the asymptotic behaviour of the integrals of the equation $y'' + Q(x)y = 0$ which makes use of recent work of my colleague G. Ascoli; this work arose partly from discussion between us regarding this new edition of my book.

(iii) The inclusion in the text and the further development of the general method of treatment of differential equations which I have called " the method of Fubini "; in the earlier edition this work appeared in an appendix.

(iv) The substantial simplification introduced in the determination of the eigenvalues of the Legendre equation in Chapter IV.

Finally, it is my pleasant duty to thank all those readers who by pointing out printing and other errors, or by suggesting possible additions and improvements, have contributed to this new edition. Among these many helpers have been my colleagues G. Ascoli, D. Graffi, L. A. MacColl, E. Persico, G. Sansone, G. Scorza-Dragoni and my assistant U. Richard, to all of whom I am particularly indebted.

F. G. T.

Turin, October 1952

Contents

I. The existence and uniqueness theorem

p. 1	1.	Résumé of some elementary theory of differential equations
4	2.	Preliminaries to the fundamental theorem
5	3.	The existence and uniqueness theorem for normal differential systems
10	4.	Additional remarks
13	5.	Circular functions
19	6.	Elliptic functions

II. The behaviour of the characteristics of a first-order equation

27	7.	Preliminary considerations
32	8.	Examples of equations with singular points
39	9.	Study of the abridged equation
45	10.	Some theorems of a general character
53	11.	The Poincaré index
55	12.	The node
63	13.	The focus and the col
74	14.	Limit cycles and relaxation oscillations
82	15.	Periodic solutions in the phase space

III. Boundary problems for linear equations of the second order

89	16.	Preliminary considerations
92	17.	A theorem of de la Vallée Poussin
96	18.	Simplifications of the given equation
98	19.	Theorems on the zeros and on the maxima and minima of integrals
101	20.	Comparison theorems and their corollaries
104	21.	The interval between successive zeros of an integral
107	22.	An important change of variable
112	23.	The oscillation theorem
117	24.	Eigenvalues and eigenfunctions

p. 119	25.	A physical interpretation
123	26.	Some properties of eigenvalues and eigenfunctions
132	27.	Connection with the theory of integral equations

IV. Asymptotic methods

139	28.	General remarks
142	29.	A general method applicable to linear differential equations
148	30.	Differential equations with stable integrals
154	31.	The case in which the coefficient of y tends to a negative limit
163	32.	Preliminaries to the asymptotic treatment of eigenvalues and of eigenfunctions
166	33.	First form of asymptotic expression for the eigenfunctions
169	34.	Asymptotic expression for the eigenvalues
174	35.	Second form of asymptotic expression for the eigenfunctions
177	36.	Equations with transition points
180	37.	The Laguerre differential equation and polynomials
186	38.	Asymptotic behaviour of the Laguerre polynomials
191	39.	The Legendre differential equation and polynomials
195	40.	An asymptotic expression for the Legendre polynomials

V. Differential equations in the complex field

202	41.	Majorizing functions
205	42.	Proof of the fundamental theorem by Cauchy's method
210	43.	General remarks on singular points of solutions of differential equations. The case of linear equations
214	44.	Investigation of the many-valuedness of integrals of a linear equation
218	45.	The case with no essential singularities
221	46.	Integration in series of equations of Fuchs' type
228	47.	Totally Fuchsian equations. The hypergeometric equation
240	48.	Preliminary remarks on points of essential singularity
245	49.	An application of the method of successive approximations
249	50.	'Asymptotic integration' of the reduced equation
253	51.	Conclusion and further comments
257	52.	Application to confluent hypergeometric functions and to Bessel functions

265	*Bibliography*
269	*Author index*
271	*General index*

I. The existence and uniqueness theorem

1. Résumé of some elementary theory of differential equations

As has been pointed out in the preface, the theory of differential equations was for many years concerned with the following problem—given a differential equation, i.e. an equation of the form*

(1) $$F(x, y, y', \ldots, y^{(n)}) = 0$$

it is required to *integrate* the equation in explicit terms, i.e. to find explicitly a function y of x and of one or more arbitrary constants $C_1, C_2, C_3, \ldots,$** which when substituted into (1) satisfies it identically, and in addition is of a *sufficient degree of generality*.

This second condition, which as stated above is vague and requires exact definition, would however be fully satisfied were it possible to write down a formula containing *all* the possible solutions of (1), as, for example, may be done in the case of an equation with *variables separable* such as

$$\frac{dy}{dx} = A(x) \cdot B(y)$$

whose solutions (usually called its *integrals*) are *all* given by the formula

$$\int \frac{dy}{B(y)} = \int A(x)\, dx + C$$

where C is an arbitrary constant. (It is assumed that $B(y) \not\equiv 0$.)

However, as it is not always possible, or at least not always easy, to determine a formula of this kind, the concept of a *general integral* of a

---

* In this book we shall not deal with equations involving *partial derivatives*. The term *differential equation* will imply an *ordinary* differential equation containing one or more unknown functions of a *single independent variable*; derivatives with respect to this variable will be denoted by *primes*.

** The function y may contain one or more integration signs if the integrations have not yet been carried out.

differential equation of the form (1) has gradually evolved in the development of modern analysis. At the present day this term denotes a function

$$(2) \qquad y = \phi(x, C_1, C_2, C_3, \ldots, C_n)$$

of the independent variable x and of n arbitrary constants $C_1, C_2, C_3, \ldots, C_n$, *the function y containing these constants in such a way that, by attributing suitable values to them, it is always possible to satisfy at a particular point $x = x_0$ "initial conditions" of the form*

$$(3) \qquad y = y_0, \; y' = y'_0, \ldots, \; y^{(n-1)} = y_0^{(n-1)}$$

for any prescribed $y_0, y'_0, y''_0, \ldots, y_0^{(n-1)}$ whatsoever.

We do not therefore exclude in any way the possible existence of further solutions of the differential equation which are not deducible from (2) by giving particular constant values to all the arbitrary constants,* but we demand only that the general integral be sufficiently *flexible* to satisfy the initial conditions (3).

Knowledge of the general integral of a differential equation is generally sufficient to enable us to solve the various problems that may arise relative to the equation—for example, that of determining the solution (or solutions) of the equation satisfying additional conditions of a different type from (3). Unfortunately the explicit determination of the general integral is not possible except in relatively rare cases and sometimes, even if possible, it may give rise to formulæ so complicated as to be of little use. Modern analysis, however, although persisting in the search for the general integral of a given equation,** prefers, as has already been pointed out, to study the main properties of its integrals *directly*, no matter whether the equation is to be studied alone or in conjunction with additional conditions of the type (3), or of another kind. We shall see this fully illustrated in the subsequent chapters.

Another concept from the elementary theory of differential equations which will be employed, in addition to that of a general integral, is that of the *equivalence of a single equation and of a system of differential equations*.

---

* Such solutions, when they exist, are called *singular integrals* of the equation. For example, *Clairaut's equation*

$$y = xy' - \phi(y')$$

has general integral $y = xc - \phi(c)$ and singular integral

$$x = \phi'(t), \; y = xt - \phi(t)$$

which is the envelope of the ∞^1 of straight lines contained in the general integral. See, for example, F. TRICOMI (79), Part II, p. 278.

** The most complete work on the less modern aspects of the theory of differential equations, i.e. the methods of obtaining the general integral, is that of A. R. FORSYTH (16).

For example, given a differential equation of order n, which for greater simplicity we shall suppose solved for the derivative of highest order, i.e. we shall suppose it written in the form

(4) $$y^{(n)} = f(x, y, y', \ldots, y^{(n-1)})$$

on putting

(5) $$y = y_1, \; y' = y_2, \; y'' = y_3, \ldots, \; y^{(n-1)} = y_n$$

we may write (4) as a system of n differential equations of the first order, viz.

(6) $$\begin{cases} y_1' = y_2 \\ \cdots\cdots \\ y_{n-1}' = y_n \\ y_n' = f(x, y_1, y_2, \ldots, y_n) \end{cases}$$

This system is a particular case of the differential system

(7) $$\begin{cases} y_1' = f_1(x, y_1, y_2, \ldots, y_n) \\ y_2' = f_2(x, y_1, y_2, \ldots, y_n) \\ \cdots\cdots\cdots\cdots\cdots \\ y_n' = f_n(x, y_1, y_2, \ldots, y_n) \end{cases}$$

Systems of the form (7) are called *normal systems* of differential equations (of the first order).

Conversely, given a system of the form (7), by differentiating each equation $(n-1)$ times with respect to x and eliminating from the n^2 equations thus obtained the $n^2 - 1$ unknowns

$$y_2, y_2', \ldots, y_2^{(n)}; \; y_3, y_3', \ldots, y_3^{(n)}; \; \ldots; \; y_n, y_n', \ldots, y_n^{(n)}$$

we obtain a single equation of order n for y_1 (solved for $y_1^{(n)}$), and similarly for $y_2, y_3, \ldots, y_n$.

Without pursuing possible easy generalizations of these last remarks—for example, their extension to a system of differential equations not all of the first order—we shall confine ourselves to the further remark that in view of (5) the initial conditions corresponding to (3) take the form

(8) $$y_1(x_0) = y_1^0, \; y_2(x_0) = y_2^0, \; \ldots, \; y_n(x_0) = y_n^0$$

where $y_1^0, y_2^0, \ldots, y_n^0$ denote arbitrary constants.

2. Preliminaries to the fundamental theorem

As first aim we wish to establish a basic fundamental theorem for the whole theory of differential equations, a theorem which asserts that, under fairly weak restrictions, a differential equation solved for the derivative of highest order possesses one and only one general integral. We thus consider, with the advantage of greater generality, a normal differential system of the form (7)*, and we shall show that under fairly wide conditions on the functions $f_1, f_2, \ldots, f_n$, there exists one and only one *solution*—i.e. one and only one *n-ple* of functions $y_1(x), \ldots, y_n(x)$ satisfying identically the system of equations and for which the initial conditions (8) are satisfied.

We demand that the functions $f_i(x, y_1, y_2, \ldots, y_n)$, $(i = 1, 2, 3, \ldots, n)$ be *continuous* with respect to the variables $x, y_1, \ldots, y_n$, and, in addition, that within suitable ranges for each of the variables $y_1, y_2, \ldots, y_n$, they satisfy a *Lipschitz condition* with respect to each variable $y_1, y_2, \ldots, y_n$. This condition is rather more restrictive than the demand of continuity but less restrictive than that of differentiability with respect to the variable concerned. In fact, a Lipschitz condition with respect to a function $F(x)$ in a certain interval (a, b) asserts that the relative increments of the function remain bounded in that interval, i.e. that there exists a positive constant A such that for any two points x_1, x_2 of the interval (a, b)

(9) $$|F(x_1) - F(x_2)| \leq A|x_1 - x_2|$$

If a function of several variables $F(x_1, x_2, \ldots, x_m)$ satisfies a Lipschitz condition *with respect to each of these variables, within suitable ranges*,** i.e. if

(10) $$|F(x_1, \ldots, x'_h, \ldots, x_m) - F(x_1, \ldots, x''_h, \ldots, x_m)| \leq A_h |x'_h - x''_h|$$

$$(h = 1, 2, \ldots, m)$$

then the function F satisfies a Lipschitz condition also *with respect to the set of variables* $x_1, x_2, \ldots, x_m$, in the sense that there exists a constant A

---

* Since a differential equation of type (4) is equivalent to a normal system of type (6) which is a particular case of system (7).

** For functions of more than one variable it may at times be important to distinguish the cases in which the Lipschitz condition is satisfied *uniformly* and those for which this is not so. The Lipschitz condition is satisfied uniformly if the constant A_h (see the following) is an absolute constant and not dependent on the other x_k. We shall suppose here and throughout the book that *the condition is satisfied uniformly*.

such that

(11) $\quad |F(x_1', x_2', \ldots, x_m') - F(x_1'', x_2'', \ldots, x_m'')|$
$$\leq A(|x_1' - x_1''| + |x_2' - x_2''| + \ldots + |x_m' - x_m''|)$$

for we have the identity

$$F(x_1', x_2', \ldots\ldots\ldots, x_m') - F(x_1'', x_2'', \ldots\ldots, x_m'')$$
$$= F(x_1', x_2', \ldots\ldots\ldots, x_m') - F(x_1'', x_2', \ldots\ldots, x_m')$$
$$+ F(x_1'', x_2', x_3', \ldots\ldots, x_m') - F(x_1'', x_2'', x_3', \ldots, x_m')$$
$$\cdots\cdots\cdots\cdots\cdots\cdots\cdots\cdots\cdots\cdots\cdots$$
$$+ F(x_1'', x_2'', \ldots, x_{m-1}'', x_m') - F(x_1'', x_2'', \ldots, x_{m-1}'', x_m'')$$

from which, by (10), follows

$$|F(x_1', x_2', \ldots, x_m') - F(x_1'', x_2'', \ldots, x_m'')|$$
$$\leq A_1|x_1' - x_1''| + A_2|x_2' - x_2''| + \ldots + A_m|x_m' - x_m''|$$

and we need only take A to be the greatest of the m constants $A_1, A_2, \ldots, A_m$ to obtain (11).

3. The existence and uniqueness theorem for normal differential systems

We can now enunciate the fundamental theorem in the following form:—
Given a differential system of the form (7) *together with the initial conditions* (8); *if two positive numbers a and b can be found such that in the domain D defined by the inequalities**

(12) $\quad\quad x_0 \leq x \leq x_0 + a;\quad\quad\quad y_1^0 - b \leq y_1 \leq y_1^0 + b$
$\quad\quad\quad\quad y_2^0 - b \leq y_2 \leq y_2^0 + b,\quad \ldots,\quad y_n^0 - b \leq y_n \leq y_n^0 + b$

the functions $f_1, f_2, \ldots, f_n$ are continuous and satisfy Lipschitz conditions in the variables $y_1, y_2, \ldots, y_n$ within these ranges, then a positive number $\delta (\leq a)$ can be determined such that in the interval $(x_0, x_0 + \delta)$ the system possesses one and only one solution satisfying the initial conditions (8).

To prove the theorem we begin by observing that if we can find in any

---
* A domain of this kind is sometimes called a *cell*, or a *hyperinterval*.

way an n-ple of functions $y_1(x), y_2(x), \ldots, y_n(x)$ satisfying the following system of n (*non-linear*) *integral equations of the Volterra type**

(13) $\qquad y_i(x) = y_i^0 + \int_{x_0}^{x} f_i[t, y_1(t), y_2(t), \ldots, y_n(t)] dt \qquad (i = 1, 2, \ldots, n)$

such an n-ple constitutes a solution of the normal system (7) which satisfies the initial conditions (8); for on differentiating (13) with respect to x we obtain the ith equation of system (7), and we need only put $x = x_0$ in (13) to verify that the function on the left-hand side assumes the value y_i^0 for $x = x_0$.

A solution of (13) for which the functions $y_i(x)$ may not be differentiable everywhere is called a solution of the differential system (7) in the sense of Carathéodory.

Now supposing that the index i takes the values $1, 2, 3, \ldots, n$, we construct an endless sequence of n-ples of functions by the recurrence formulæ

(14) $\qquad y_i^0(x) = y_i^0, \; y_i^{(m+1)}(x) = y_i^0 + \int_{x_0}^{x} f_i[t, y_1^{(m)}(t), y_2^{(m)}(t), \ldots, y_n^{(m)}(t)] dt$
$\qquad\qquad\qquad\qquad\qquad\qquad\qquad\qquad\qquad (m = 0, 1, 2, \ldots)$

(Note that here the upper suffices do not denote derivatives.) We shall show first that given a positive number M such that throughout the entire domain D the continuous functions f_i satisfy the inequalities

(15) $\qquad\qquad |f_i(x, y_1, y_2, \ldots, y_n)| \leq M \qquad (i = 1, 2, \ldots, n)$

then for values of x in the interval $(x_0, x_0 + \delta)$ where δ is the smaller of the two numbers a and b/M (i.e. $\delta = \min. (a, b/M)$), the functions $y_i^{(1)}(x)$, $y_i^{(2)}(x), \ldots$, are always contained between $y_i^0 - b$ and $y_i^0 + b$.

We have, recursively,

$$|y_i^{(1)}(x) - y_i^0| \leq \int_{x_0}^{x} |f_i(t, y_1^0, y_2^0, \ldots, y_n^0)| dt \leq M(x - x_0) \leq M\delta \leq b$$

$$|y_i^{(2)}(x) - y_i^0| \leq \int_{x_0}^{x} |f_i[t, y_1^{(1)}(t), y_2^{(1)}(t), \ldots, y_n^{(1)}(t)]| dt \leq M(x - x_0) \leq b$$

. . . .

---

* An *integral equation* is one in which the unknown function (or functions) appears (throughout the equation or in part) under the sign of integration—see, for example, F. TRICOMI (82). In this book we shall not deal systematically with such equations although they are closely connected with many parts of the theory of differential equations; but it is impossible to avoid some consideration of integral equations, particularly of the most elementary type (the Volterra equations in which the range of integration of the integrals has lower limit some fixed constant x_0 and upper limit the independent variable x). We shall derive the properties of these equations as required.

§3] The existence and uniqueness theorem

This is of importance because it ensures—provided that x remains in the interval $(x_0, x_0 + \delta)$—that the functions f_i having for argument y some one n-ple $(y_1^{(m)}, y_2^{(m)}, \ldots, y_n^{(m)})$ as defined above satisfy the hypotheses of the theorem, and, in particular, the Lipschitz conditions.

We now show that *the successive n-ples defined above tend uniformly to an n-ple limit*

(16) $$Y_1(x), Y_2(x), \ldots, Y_n(x)$$

i.e. given a positive number ε as small as we please, we can always find an m_0 such that, provided x lies between x_0 and $x_0 + \delta$ and i between 1 and n, we have for $m > m_0$

(17) $$|Y_i(x) - y_i^{(m)}(x)| < \varepsilon$$

For employing the first of the inequalities established above, viz.

$$|y_i^{(1)}(x) - y_i^0| \leq M(x - x_0)$$

and applying to the functions f_i the Lipschitz inequality (11), we can write

$$|y_i^{(2)}(x) - y_i^{(1)}(x)| \leq \int_{x_0}^{x} |f_i[t, y_1^{(1)}(t), y_2^{(1)}(t), \ldots, y_n^{(1)}(t)] - f_i[t, y_1^0, y_2^0, \ldots, y_n^0]| dt$$

$$\leq A_i \int_{x_0}^{x} (|y_1^{(1)}(t) - y_1^0| + |y_2^{(1)}(t) - y_2^0| + \ldots + |y_n^{(1)}(t) - y_n^0|) dt$$

$$\leq A_i n M \int_{x_0}^{x} (t - x_0) dt = A_i n M \frac{(x - x_0)^2}{2!}$$

where A_i is a suitable positive constant. Similarly

$$|y_i^{(3)}(x) - y_i^{(2)}(x)|$$

$$\leq \int_{x_0}^{x} |f_i[t, y_1^{(2)}(t), y_2^{(2)}(t), \ldots, y_n^{(2)}(t)] - f_i[t, y_1^{(1)}(t), y_2^{(1)}(t), \ldots, y_n^{(1)}(t)]| dt$$

$$\leq A_i \int_{x_0}^{x} (|y_1^{(2)}(t) - y_1^{(1)}(t)| + |y_2^{(2)}(t) - y_2^{(1)}(t)| + \ldots + |y_n^{(2)}(t) - y_n^{(1)}(t)|) dt$$

$$\leq A_i n \cdot A_i n M \int_{x_0}^{x} \frac{(t - x_0)^2}{2!} dt = (A_i n)^2 M \frac{(x - x_0)^3}{3!}$$

and so on. Thus, in general,

$$|y_i^{(m)}(x) - y_i^{(m-1)}(x)| \leq (A_i n)^{m-1} M \frac{(x - x_0)^m}{m!} \leq (A_i n)^{m-1} M \frac{\delta^m}{m!}$$

or

(18) $$|y_i^{(m)}(x) - y_i^{(m-1)}(x)| \leq \frac{M}{nA_i} \frac{(nA_i \delta)^m}{m!}$$

From this it follows that the series

(19) $\quad y_i^0 + [y_i^{(1)}(x) - y_i^0] + [y_i^{(2)}(x) - y_i^{(1)}(x)] + \ldots + [y_i^{(m)}(x) - y_i^{(m-1)}(x)] + \ldots$

is majorized by the convergent series of positive constants

$$\frac{M}{nA_i} \sum_{m=0}^{\infty} \frac{(nA_i\delta)^m}{m!} = \frac{M}{nA_i} e^{nA_i\delta}$$

and is therefore absolutely and uniformly convergent in the interval $(x_0, x_0 + \delta)$. Hence, since its sum to $(m+1)$ terms is $y_i^{(m)}(x)$, it follows that, for any x in $(x_0, x_0 + \delta)$ and for $m >$ a suitable $m_{0,\,i}$,

$$|Y_i(x) - y_i^{(m)}(x)| < \varepsilon$$

where $Y_i(x)$ denotes the sum of the series (19), i.e. the limit of $y_i^{(m)}(x)$ as $m \to \infty$.

We need now only take m_0 as the greatest of the numbers $m_{0,\,1}, m_{0,\,2}, \ldots, m_{0,\,n}$, to establish the result enunciated above.

The last step now to be done, consists in showing that the n-ple limit (16) satisfies the system of integral equations (13), i.e. that the relations

(20) $\quad Y_i(x) = y_i^0 + \int_{x_0}^{x} f_i[t, Y_1(t), Y_2(t), \ldots, Y_n(t)] dt \quad (i = 1, 2, \ldots, n)$

are identities.

To do this it is convenient to put

$$Y_i(x) - y_i^{(m)}(x) = R_i^{(m)}(x)$$

The recurrence formulæ (14) now yield

$Y_i(x) - R_i^{(m+1)}(x)$

$\quad = y_i^0 + \int_{x_0}^{x} f_i[t, Y_1(t) - R_1^{(m)}(t), Y_2(t) - R_2^{(m)}(t), \ldots, Y_n(t) - R_n^{(m)}(t)] dt$

from which, by rearranging and subtracting from both sides of this equation

$$\int_{x_0}^{x} f_i[t, Y_1(t), Y_2(t), \ldots, Y_n(t)] dt$$

follows the identity

$Y_i(x) - y_i^0 - \int_{x_0}^{x} f_i[t, Y_1(t), Y_2(t), \ldots, Y_n(t)] dt = R_i^{(m+1)}(x)$

$\quad + \int_{x_0}^{x} \{f_i[t, Y_1(t) - R_1^{(m)}(t), \ldots, Y_n(t) - R_n^{(m)}(t)] - f_i[t, Y_1(t), \ldots, Y_n(t)]\} dt$

But in virtue of the Lipschitz conditions and (17) we have

$$|f_i[t, Y_1(t)-R_1^{(m)}(t), \ldots, Y_n(t)-R_n^{(m)}(t)] - f_i[t, Y_1(t), \ldots, Y_n(t)]|$$
$$\leq A_i(|R_1^{(m)}(t)| + \ldots + |R_n^{(m)}(t)|) < nA_i\varepsilon$$

and thus, by taking absolute values on both sides in the preceding identity, we deduce

$$\left|Y_i(x) - y_i^0 - \int_{x_0}^x f_i[t, Y_1(t), \ldots, Y_n(t)]dt\right| < |R_i^{(m+1)}(x)| + nA_i\varepsilon \int_{x_0}^x dt < (1+nA_i\delta)\varepsilon$$

In these inequalities the first term is some non-negative number, while the last term can be made as small as we please by suitably choosing ε; thus the first term must be equal to zero, i.e. (20) is satisfied identically.

Finally, we show that the *n-ple* (16) constitutes the unique solution of the system of integral equations (13).

For suppose the *n*-ple

(21) $$Y_1^*(x), Y_2^*(x), \ldots, Y_n^*(x)$$

is also a solution of system (13). Clearly the functions $Y_i^*(x)$ must be continuous because so is the right-hand side of (13); also $Y_i^*(x_0) = y_i^0$. Thus replacing (if necessary) the interval $(x_0, x_0 + \delta)$ by a more restricted interval $(x_0, x_0 + \delta')$ we may suppose that

$$|Y_i^*(x) - y_i^0| \leq b \qquad (i = 1, 2, \ldots, n)$$

Now employing (15) and the equality

$$Y_i^*(x) = y_i^0 + \int_{x_0}^x f_i[t, Y_1^*(t), Y_2^*(t), \ldots, Y_n^*(t)]dt$$

we deduce that

$$|Y_i^*(x) - y_i^0| \leq M(x - x_0)$$

On the other hand, by subtracting term by term from the preceding equality the recursive formula (14) we obtain

$$Y_i^*(x) - y_i^{(m+1)}(x) = \int_{x_0}^x \{f_i[t, Y_1^*(t), \ldots, Y_n^*(t)] - f_i[t, y_1^{(m)}(t), \ldots, y_n^{(m)}(t)]\}dt$$

from which, by taking absolute values on both sides and applying the Lipschitz inequality, it follows that

$$|Y_i^*(x) - y_i^{(m+1)}(x)| \leq A_i \int_{x_0}^x \{|Y_1^*(t) - y_1^{(m)}(t)| + \ldots + |Y_n^*(t) - y_n^{(m)}(t)|\}dt$$

Thus, by putting $m = 0$ and using the penultimate inequality, we derive

$$|Y_i^*(x) - y_i^{(1)}(x)| \leq A_i n M \int_{x_0}^{x} (t-x_0)dt = A_i n M \frac{(x-x_0)^2}{2!}$$

whence, by putting $m = 1$ and using the result just obtained,

$$|Y_i^*(x) - y_i^{(2)}(x)| \leq A_i \int_{x_0}^{x} \{|Y_1^*(t) - y_1^{(1)}(t)| + \ldots + |Y_n^*(t) - y_n^{(1)}(t)|\}dt$$

$$\leq A_i n \cdot A_i n M \int_{x_0}^{x} \frac{(t-x_0)^2}{2!} dt = (A_i n)^2 M \frac{(x-x_0)^3}{3!}$$

and so on. Thus in general we obtain

$$|Y_i^*(x) - y_i^{(m)}(x)| \leq \frac{M}{nA_i} \frac{[nA_i(x-x_0)]^{m+1}}{(m+1)!}$$

from which, since the general term in a convergent series must tend to zero, it follows that

$$\lim_{m \to \infty} [Y_i^*(x) - y_i^{(m)}(x)] = 0$$

i.e.

$$Y_i^*(x) = \lim_{m \to \infty} y_i^{(m)}(x) = Y_i(x) \qquad (i = 1, 2, \ldots, n)$$

showing that *the n-ple (21) is identical with the n-ple (16)*.

This completes the proof of the theorem of existence and uniqueness.

4. Additional remarks

It is repetitive to stress again the importance of the result just obtained, but it provides the rigorous basis of the whole theory of differential equations. It is not only the result itself which is important but also the *method of successive approximations** which establishes it, for this method may be extended in various ways; also it furnishes a process of calculation which,

---

* With the method of successive approximations is usually associated the name of the French mathematician E. Picard (1856–1941) who emphasized its importance. The method however was used some years earlier by G. Peano (1858–1932) and still earlier by J. Liouville (1809–1882) and others.

although generally laborious, nevertheless effectively produces as close an approximation as desired to the solution whose existence and uniqueness have just been established. We are therefore dealing with a *constructive method* of proof. Several examples which we shall presently discuss will illustrate this.

We can, in some cases, widen the scope of the result derived in the preceding section. First, the behaviour which is described *to the right* of the point x_0 may be repeated *to the left*; thus if the hypotheses of the theorem are satisfied within the domain D', in place of the domain D defined by the inequalities (12), where within D' the previous limits on y remain valid while x varies in the interval $(x_0 - a', x_0)$ where $a' > 0$, then the theorem is valid in the interval $(x_0 - \delta', x_0)$ where δ' is the smaller of the two positive numbers a' and b/M. Hence if the hypotheses are actually satisfied for $x_0 - a' \leq x \leq x_0 + a$, the existence and uniqueness are established in a two-sided interval about x_0, viz.

$$x_0 - \delta' \leq x \leq x_0 + \delta$$

A further important point to be noted is that the n functions $Y_i(x)$ constituting the solution described by the fundamental theorem may be considered as the sums of the *uniformly* convergent series (19). Thus *in the case in which the functions f_i are continuous functions of one or more parameters $\lambda, \mu, \ldots, y_i^{(m)}(x)$* are evidently continuous functions of the same parameters; hence *the functions $Y_i(x)$ must be continuous functions of $\lambda, \mu, \ldots$*, as the sum of a uniformly convergent series of continuous functions is also a continuous function.

Therefore with suitable adjustments to the proof, as indicated above, the fundamental theorem may be stated as in the following paragraph, in which the initial values of the y_i are now allowed to vary within suitable intervals and consequently in which *the functions $Y_i(x)$ are continuous functions of the initial values y_i^0*. This turns out to be very useful.

The most general statement of the fundamental theorem is as follows:

Given a normal differential system of the form (7) *and an n-ple of constants $\beta_1, \beta_2, \ldots, \beta_n$; if two positive constants a and b can be determined such that within the domain D defined by the inequalities*

(22) $$x_0 \leq x \leq x_0 + a$$

$$\beta_1 - b \leq y_1 \leq \beta_1 + b, \quad \beta_2 - b \leq y_2 \leq \beta_2 + b, \quad \ldots, \quad \beta_n - b \leq y_n \leq \beta_n + b$$

the functions $f_1, f_2, \ldots, f_n$ are continuous and satisfy Lipschitz conditions in the variables $y_1, y_2, \ldots, y_n$ within the given ranges, then a positive number δ

may be found ($\delta \leq a$) such that, given any set $y_1^0, y_2^0, y_3^0, \ldots, y_n^0$ of initial values satisfying

(23) $$|y_i^0 - \beta_i| \leq \frac{b}{2} \qquad (i = 1, 2, \ldots, n)$$

the differential system (7) *possesses one and only one solution which satisfies the initial conditions* (8) *in the interval* $(x_0, x_0 + \delta)$.

If in the domain D the functions f_i are all in absolute value less than M, δ may be taken equal to the minimum of a and $b/(4M)$.

This more general form of the fundamental theorem can be established by suitably amending the proof given for the weaker theorem. See, for example, Sansone (47), chapter I. This reference may be consulted also in regard to the following problem which we can merely mention here.

What happens if the functions f_i are continuous but fail to satisfy the Lipschitz conditions in the variables y_h within the given ranges?

It is fairly evident that even under these more general hypotheses *the existence part of the theorem remains valid,* but in general *the uniqueness part is no longer true,* as will be shown in an example we shall consider at the beginning of the following chapter.

In view of the character of this book we cannot consider in detail this important extension of the existence theorem*—due to Peano—which requires less simple expansions than we have used, and which in fact belongs naturally to a field that we cannot even touch upon. This extension is concerned with existence and uniqueness theorems *in the large,* i.e. with theorems of this kind valid not only in a sufficiently small region about a certain point, but valid throughout a whole domain in which the equation satisfies certain conditions.

We note finally, in connection with these last points, that the solution whose existence is established by the fundamental theorem in the interval $(x_0, x_0 + \delta)$ may, in general, be *extended* into a further interval $(x_0 + \delta, x_0 + \delta + \delta_1)$ by applying the same theorem to the point $x_1 = x_0 + \delta$ with initial values equal to the final values of the solution obtained immediately previously, and so on. It must not however be assumed that this process can be continued until the conditions on the functions f_i are no longer satisfied, i.e. it must not be assumed that if the set of numbers $x_0 + \delta$, $x_0 + \delta + \delta_1$, $x_0 + \delta + \delta_1 + \delta_2, \ldots$, has finite upper bound ξ the point $x = \xi$ is necessarily a point in which at least one of the functions f_i ceases to satisfy (at least for some set of values of the y_i's) the conditions demanded in the fundamental theorem.**

---

* G. Sansone (47), chap. VIII. Further references are also given here.
** This is however the case, as we shall see in § 43, for *linear* equations.

To illustrate this point, it is sufficient to consider the simple example of the single (non-linear) equation of the first order

(24) $$y' + y^2 = 0$$

whose general integral is

$$y = \frac{1}{x+c}$$

Clearly the integral

(25) $$y = \frac{1}{x - x_0 + 1/y_0}$$

which for $x = x_0$ takes the value y_0, cannot be continued beyond the point $x = x_0 - 1/y_0$ (where y becomes infinite) despite the fact that at that point there is no singularity of (24)—in which equation, in fact, x does not appear explicitly.

5. Circular functions

It is instructive from the point of view of the general theory, to study the differential system

(26) $$y_1' = y_2, \qquad y_2' = -y_1$$

together with the initial conditions

(27) $$y_1(0) = 0, \qquad y_2(0) = 1$$

Disregarding meantime the familiar properties of its solutions which are $y_1(x) = \sin x$, $y_2(x) = \cos x$, we try to *read off* from equations (26) and (27) the principal properties of the solutions of the system.

We first investigate how the method of successive approximations used in the proof of the fundamental theorem applies in this case. As in this example,

$$f_1(x, y_1, y_2) = y_2, \qquad f_2(x, y_1, y_2) = -y_1, \qquad x_0 = 0, \qquad y_1^0 = 0, y_2^0 = 1$$

the recurrence formulæ for the successive approximations take the form

$$y_1^{(m+1)}(x) = \int_0^x y_2^{(m)}(t)dt, \qquad y_2^{(m+1)}(x) = 1 - \int_0^x y_1^{(m)}(t)dt$$

and elementary calculations yield

$$\begin{cases} y_1^{(1)} = x \\ y_2^{(1)} = 1 \end{cases} \begin{cases} y_1^{(2)} = x \\ y_2^{(2)} = 1 - \dfrac{x^2}{2!} \end{cases} \begin{cases} y_1^{(3)} = x - \dfrac{x^3}{3!} \\ y_2^{(3)} = 1 - \dfrac{x^2}{2!} \end{cases} \begin{cases} y_1^{(4)} = x - \dfrac{x^3}{3!} \\ y_2^{(4)} = 1 - \dfrac{x^2}{2!} + \dfrac{x^4}{4!} \end{cases}$$

and so on.

Hence successive approximations give successive partial sums of the series (but here with each sum repeated twice over)

(28) $\quad y_1(x) = x - \dfrac{x^3}{3!} + \dfrac{x^5}{5!} - \ldots, \quad y_2(x) = 1 - \dfrac{x^2}{2!} + \dfrac{x^4}{4!} - \ldots$

which are everywhere convergent and which are the expansions as power series of the functions $\sin x$ and $\cos x$. In other words, in this actual case, the method of successive approximations yields the power-series expansions of the two solution functions, and as these power series are convergent for all x we deduce that the two solutions in question are bounded.

Several properties—for example, that $\sin x$ and $\cos x$ lie between -1 and $+1$, or that they are periodic of period 2π—may be deduced* fairly easily from the series (28); but it is of interest, as has already been suggested, to look for a method of reading off these properties *directly* from the differential equations.

We observe initially that by multiplying the first of equations (26) by $2y_1$, the second by $2y_2$ and adding, we obtain

$$2y_1 y_1' + 2y_2 y_2' = 0$$

But the left-hand side is the derivative of $y_1^2 + y_2^2$, and therefore

$$y_1^2 + y_2^2 = \text{constant}$$

On account of the initial conditions (27) this becomes

(29) $\quad y_1^2 + y_2^2 = 1$

From equation (29) follow several corollaries. First,

$$|y_1| \leq 1, \quad |y_2| \leq 1$$

Second, y_1 and y_2 *can never vanish simultaneously; also their zeros are all*

---

* For proofs, see K. KNOPP (67).

simple, for if y_1 and y'_1 vanish simultaneously* at a certain point (as happens at every multiple zero of y_1) then, by the first equation of (26), $y_2 = 0$ at that point.

But do the functions y_1 and y_2 actually vanish at any points?

For the function $y_1(x)$ this is evidently so, since the series which appears in the first equation of (28) vanishes at least for $x = 0$. For the function $y_2(x)$ there may in fact be dubiety, but this is easily removed by a *reductio ad absurdum* argument as follows.

If y_2 never vanishes we may suppose y_2 always positive for $x > 0$ since $y_2(0) = 1$; then $y_1(0) = 0$ and $y'_1 = y_2 > 0$; thus the function y_1 is always positive for $x > 0$, so that for $x \geq \varepsilon > 0$, $y_1(x) \geq \eta$ where $y_1(\varepsilon) = \eta$. But since $y'_2 = -y_1$, the preceding inequality obviously implies

$$|y_2(x) - y_2(\varepsilon)| \geq \eta(x - \varepsilon) \qquad (x \geq \varepsilon) $$

which is incompatible with the fact that $y_2(x)$ is always bounded—and the result follows.

The differential system (26), and in fact all systems in which the independent variable x does not appear explicitly, remains invariant under a transformation of the form $x = \xi + $ constant; further, (26) remains unaltered (in view of the particular form of its equations) by an interchange of y_1 and y_2, provided that a sign change in either of these variables is introduced. We now make a change of origin to the first positive zero (which we call α) of the function $y_2(x)$ at which point we have, by (29),**

$$y_1(\alpha) = 1, \qquad y_2(\alpha) = 0$$

the change of variable being given by

(30) $$x = \xi + \alpha$$

Now putting

(31) $$y_1(x) = z_2(\xi), \qquad y_2(x) = -z_1(\xi)$$

the functions $z_1(\xi)$ and $z_2(\xi)$ satisfy a differential system similar to system (26), viz.

$$z'_1 = z_2, \qquad z'_2 = -z_1$$

---

* In general, n functions $y_1, y_2, \ldots, y_n$, not all identically zero, which satisfy a normal system of type (7) and are such that when $y_1 = y_2 = \ldots = y_n = 0$ the n functions $f_1, f_2, \ldots, f_n$ vanish (independently of the value of x), *cannot vanish simultaneously at some particular point*; because assuming this point as initial point x_0, the system and the initial conditions are then satisfied by $y_1 \equiv y_2 \equiv \ldots \equiv y_n \equiv 0$, and therefore by the uniqueness theorem there exists no n-ple of functions $y_1, y_2, \ldots, y_n$ as was assumed initially.

** Clearly $y_1(\alpha) = +1$ since the function $y_1(x)$ is increasing in the interval $(0, \alpha)$ in which $y_2(x)$ is positive. (This follows from the first equation (26).)

together with the initial conditions

$$z_1(0) = -y_2(\alpha) = 0, \qquad z_2(0) = y_1(\alpha) = 1$$

These initial conditions are therefore exactly analogous to the conditions (27) associated with system (26). Hence by the uniqueness theorem it necessarily follows that $z_1(\xi) = y_1(\xi)$, $z_2(\xi) = y_2(\xi)$; these imply

$$-y_2(x) = y_1(\xi), \qquad y_1(x) = y_2(\xi)$$

from which follow the formulæ

(32) $$y_1(\xi+\alpha) = y_2(\xi), \qquad y_2(\xi+\alpha) = -y_1(\xi)$$

From the results immediately above we deduce first that *the functions $y_1(x)$ and $y_2(x)$ are periodic with period 4α*; for it follows that

(33) $\quad y_1(x+2\alpha) = y_2(x+\alpha) = -y_1(x), \qquad y_2(x+2\alpha) = -y_1(x+\alpha) = -y_2(x)$

and successively that

$$y_1(x+4\alpha) = -y_1(x+2\alpha) = y_1(x), \qquad y_2(x+4\alpha) = -y_2(x+2\alpha) = y_2(x)$$

Further, from formulæ (32) follow the relations

$$y_1(2\alpha) = y_2(\alpha) = 0, \quad y_2(3\alpha) = -y_1(2\alpha) = 0, \quad y_1(4\alpha) = y_2(3\alpha) = 0, \quad \ldots$$

Thus *the even multiples of α are zeros of $y_1(x)$ while the odd multiples of the same α are zeros of $y_2(x)$*; and such multiples are *the only zeros of the two functions*.

If α is the *first* positive zero of the *even* function $y_2(x)$ this function is non-zero in the interval $-\alpha < x < \alpha$; and so, in virtue of the second equation of (33), also in the interval $\alpha < x < 3\alpha$; hence the only possible zeros of $y_2(x)$ in the interval $(-\alpha, 3\alpha)$, which on account of periodicity is all we need consider, are the points $x = -\alpha$, $x = \alpha$, and $x = 3\alpha$.

Also, the only possible zeros of the function $y_1(x) = y_2(x-\alpha)$ in the interval $(0, 4\alpha)$ are the points $x = 0$, $x = 2\alpha$, and $x = 4\alpha$. Equation (26) implies that when y_2 is positive (negative) y_1 is increasing (decreasing), while when y_1 is positive (negative) y_2 is decreasing (increasing); also that when one of these two functions vanishes the other passes through its maximum value $+1$ or its minimum value -1. We therefore deduce that the functions $y_1(x)$ and $y_2(x)$ have the familiar oscillatory forms of $\sin x$ and $\cos x$ respectively, provided that the constant α is identified with $\pi/2$.

Such identification is justified by showing from the results previously obtained that the area A of a circle of unit radius equals 2α. Now

$$\tfrac{1}{4}A = \int_0^1 \sqrt{1-x^2}\, dx$$

so that on integration by parts

$$\tfrac{1}{4}A = [x\sqrt{1-x^2}]_0^1 - \int_0^1 \frac{1-x^2-1}{\sqrt{1-x^2}}\, dx = \int_0^1 \frac{dx}{\sqrt{1-x^2}} - \tfrac{1}{4}A$$

and therefore

$$\tfrac{1}{2}A = \int_0^1 \frac{dx}{\sqrt{1-x^2}}$$

On making the change of variable

$$x = y_1(\xi)$$

and using the first equation in (26), we have

$$dx = y_1'(\xi)d\xi = y_2(\xi)d\xi$$

so that

$$\tfrac{1}{2}A = \int_0^\alpha \frac{y_2(\xi)}{\sqrt{1-y_1^2(\xi)}}\, d\xi$$

But from (29), since $\sqrt{1-y_1^2(\xi)} = y_2(\xi)$,* this last integrand reduces to 1 and so $\tfrac{1}{2}A = \alpha$, i.e. $A = 2\alpha$.

Lastly we show how, by using (26), we can easily deduce the addition theorems for the functions $y_1(x)$ and $y_2(x)$, viz.

(34)
$$y_1(\xi+h) = y_1(\xi)y_2(h) + y_2(\xi)y_1(h)$$
$$y_2(\xi+h) = y_2(\xi)y_2(h) - y_1(\xi)y_1(h)$$

These formulæ, as is immediately obvious, are the well-known addition formulæ for sines and cosines, the basic formulæ in the theory of circular functions.

We make use of the property, already emphasized, that the system is invariant under a substitution of the form $x = \xi + h$; also of the further property that for a normal *linear homogeneous system* such as (26) and, in general, of the form

(35) $\qquad y_i' = a_{i1}(x)y_1 + a_{i2}(x)y_2 + \ldots + a_{in}(x)y_n \qquad (i = 1, 2, \ldots, n)$

* Equality in *magnitude and sign* as $y_2(\xi)$ is positive within the interval $(0, \alpha)$, as is also the radical.

where $a_{i1}(x), a_{i2}(x), \ldots, a_{in}(x)$ denote (in general) arbitrary functions of the independent variable x, every solution is given by

$$(36) \qquad Y_i = C_1 Y_i^{(1)} + C_2 Y_i^{(2)} + \ldots + C_n Y_i^{(n)} \qquad (i = 1, 2, \ldots, n)$$

where $Y_i^{(1)}(x), Y_i^{(2)}(x), \ldots, Y_i^{(n)}(x)$ are n *linearly independent** solutions of (35) and $C_1, C_2, C_3, \ldots, C_n$ denote n arbitrary constants.

The Y_i's which appear in (36) evidently satisfy system (35) since, for all values of x,

$$Y_i' = \sum_{k=1}^{n} C_k \frac{d}{dx} Y_i^{(k)}$$

$$= \sum_{k=1}^{n} C_k [a_{i1}(x) Y_1^{(k)} + a_{i2}(x) Y_2^{(k)} + \ldots + a_{in}(x) Y_n^{(k)}]$$

$$= a_{i1}(x) \sum_{k=1}^{n} C_k Y_1^{(k)} + a_{i2}(x) \sum_{k=1}^{n} C_k Y_2^{(k)} + \ldots + a_{in}(x) \sum_{k=1}^{n} C_k Y_n^{(k)}$$

$$= a_{i1}(x) Y_1 + a_{i2}(x) Y_2 + \ldots + a_{in}(x) Y_n \qquad (i = 1, 2, \ldots, n)$$

In addition these functions Y_i can be made to satisfy the arbitrary initial conditions

$$Y_1(x_0) = y_1^0, \qquad Y_2(x_0) = y_2^0, \qquad \ldots, \qquad Y_n(x_0) = y_n^0$$

where $y_1^0, y_2^0, \ldots, y_n^0$ denote any constants whatsoever, since it is only necessary to choose $C_1, C_2, \ldots, C_n$ in such a way that the following system of n linear equations in n unknowns is satisfied (the determinant formed by the coefficients being non-zero**)

$$C_1 Y_i^{(1)}(x_0) + C_2 Y_i^{(2)}(x_0) + \ldots + C_n Y_i^{(n)}(x_0) = y_i^0 \qquad (i = 1, 2, \ldots, n)$$

It follows therefore that the functions

$$y_1(\xi + h) \qquad \text{and} \qquad y_2(\xi + h)$$

considered as functions of ξ must satisfy (26); they may therefore be written

---

* A set of functions is said to be *linearly independent* if no linear combination of them is identically zero.

** For if the determinant were zero there would exist a set of n constants $\mu_1, \mu_2, \ldots, \mu_n$, not all zero, such that

$$\mu_1 Y_i^{(1)} + \mu_2 Y_i^{(2)} + \ldots + \mu_n Y_i^{(n)} \equiv 0 \quad (i = 1, 2, \ldots, n),$$

contrary to the hypothesis that the n solutions considered are linearly independent.

as *linear combinations of any two linearly independent solutions of the same system,* for example, as linear combinations of either pair

$$\begin{cases} y_1(\xi) \\ y_2(\xi) \end{cases} \qquad \begin{cases} y_1(\xi+\alpha) = y_2(\xi) \\ y_2(\xi+\alpha) = -y_1(\xi) \end{cases}$$

Thus

(37) $\qquad y_1(\xi+h) = c_1 y_1(\xi) + c_2 y_2(\xi), \qquad y_2(\xi+h) = c_1 y_2(\xi) - c_2 y_1(\xi),$

where c_1 and c_2 denote two quantities independent of ξ which can be immediately determined on putting $\xi = 0$. This gives

$$y_1(h) = c_2, \qquad y_2(h) = c_1$$

We need only substitute these values into (37) to obtain (34).

We note in conclusion that in view of (29) the differential equations from which we started may be put into the "variables separable" forms

$$\frac{dy_1}{dx} = \sqrt{1-y_1^2}, \qquad \frac{dy_2}{dx} = -\sqrt{1-y_2^2}$$

Consequently the functions y_1 and y_2 may be studied as *the inverse functions of the two indefinite integrals*

$$x = \int \frac{dy_1}{\sqrt{1-y_1^2}}, \qquad x = -\int \frac{dy_2}{\sqrt{1-y_2^2}}$$

This approach however introduces the considerable difficulty of proving that such functions are in fact *uniform*.

6. Elliptic functions

We now apply methods similar to those used in the preceding section to the differential system

(38) $\qquad y_1' = y_2 y_3, \qquad y_2' = -y_1 y_3, \qquad y_3' = -k^2 y_1 y_2$

with initial conditions

(39) $\qquad y_1(0) = 0, \qquad y_2(0) = 1, \qquad y_3(0) = 1,$

where k denotes a constant lying between 0 and 1. This system no longer leads to elementary functions but to one of the most important classes of non-elementary transcendental functions, viz. *the Jacobi elliptic functions*.

In this case where $n = 3$ and

$$f_1(x, y_1, y_2, y_3) = y_2 y_3, \qquad f_2(x, y_1, y_2, y_3) = -y_1 y_3$$
$$f_3(x, y_1, y_2, y_3) = -k^2 y_1 y_2$$
$$x_0 = 0, \qquad y_1^0 = 0, \qquad y_2^0 = 1, \qquad y_3^0 = 1,$$

the recurrence relations between successive approximations take the form

$$y_1^{(m+1)}(x) = \int_0^x y_2^{(m)}(t) y_3^{(m)}(t) dt, \qquad y_2^{(m+1)}(x) = 1 - \int_0^x y_1^{(m)}(t) y_3^{(m)}(t) dt$$

$$y_3^{(m+1)}(x) = 1 - k^2 \int_0^x y_1^{(m)}(t) y_2^{(m)}(t) dt$$

Easy calculation gives

$$\begin{cases} y_1^{(1)} = x \\ y_2^{(1)} = 1 \\ y_3^{(1)} = 1 \end{cases} \quad \begin{cases} y_1^{(2)} = x \\ y_2^{(2)} = 1 - \dfrac{x^2}{2!} \\ y_3^{(2)} = 1 - k^2 \dfrac{x^2}{2!} \end{cases} \quad \begin{cases} y_1^{(3)} = x - (1+k^2)\dfrac{x^3}{3!} + 6k^2 \dfrac{x^5}{5!} \\ y_2^{(3)} = 1 - \dfrac{x^2}{2!} + 3k^2 \dfrac{x^4}{4!} \\ y_3^{(3)} = 1 - k^2 \dfrac{x^2}{2!} + 3k^2 \dfrac{x^4}{4!} \end{cases}$$

and successively

$$\begin{cases} y_1^{(4)} = x - (1+k^2)\dfrac{x^3}{3!} + 12k^2 \dfrac{x^5}{5!} + \dots \\ y_2^{(4)} = 1 - \dfrac{x^2}{2!} + (1+4k^2)\dfrac{x^4}{4!} + \dots \\ y_3^{(4)} = 1 - k^2 \dfrac{x^2}{2!} + k^2(4+k^2)\dfrac{x^4}{4!} + \dots \end{cases}$$

$$\begin{cases} y_1^{(5)} = x - (1+k^2)\dfrac{x^3}{3!} + (1+14k^2+k^4)\dfrac{x^5}{5!} + \dots \\ y_2^{(5)} = 1 - \dfrac{x^2}{2!} + (1+4k^2)\dfrac{x^4}{4!} - k^2(42+15k^2)\dfrac{x^6}{6!} + \dots \\ y_3^{(5)} = 1 - k^2 \dfrac{x^2}{2!} + k^2(4+k^2)\dfrac{x^4}{4!} - k^2(15+42k^2)\dfrac{x^6}{6!} + \dots \end{cases}$$

where the dots denote terms (finite in number) of higher degree in x than those written explicitly.

It happens that the coefficients of the various powers of x ultimately become *stable*, i.e. they ultimately remain unchanged in successive approximations, and these stable coefficients form three power series in x which, at least within a sufficiently small neighbourhood of the origin, represent the functions y_1, y_2, y_3, because the differences between the three functions $y_1^{(m)}$, $y_2^{(m)}$, $y_3^{(m)}$ of the mth approximations and the three corresponding partial sums of these series tend to zero as $m \to \infty$: the three series are

$$(40) \quad \begin{cases} y_1(x) = x - (1+k^2)\frac{x^3}{3!} + (1+14k^2+k^4)\frac{x^5}{5!} - \ldots \\ y_2(x) = 1 - \frac{x^2}{2!} + (1+4k^2)\frac{x^4}{4!} - (1+44k^2+16k^4)\frac{x^6}{6!} + \ldots \\ y_3(x) = 1 - k^2\frac{x^2}{2!} + k^2(4+k^2)\frac{x^4}{4!} - k^2(16+44k^2+k^4)\frac{x^6}{6!} + \ldots \end{cases}$$

However the series obtained here, in addition to being rather more complicated than (28), are *not* always convergent (they have finite radius of convergence) and it is therefore even more convenient now than in the previous example to derive *directly* from the equations (38) the properties of these functions y_1, y_2, y_3, which are usually denoted by the symbols sn x, cn x, dn x respectively.

In system (38) we multiply the first equation by $2y_1$, the second by $2y_2$ and add together; we again multiply the first equation by $2k^2 y_1$, the third by $2y_3$ and add together; we thus obtain the equations

$$2y_1 y_1' + 2y_2 y_2' = \frac{d}{dx}(y_1^2 + y_2^2) = 0$$

$$2k^2 y_1 y_1' + 2y_3 y_3' = \frac{d}{dx}(k^2 y_1^2 + y_3^2) = 0$$

whose integrals are

$$y_1^2 + y_2^2 = \text{constant}, \qquad k^2 y_1^2 + y_3^2 = \text{constant}$$

On taking account of the initial conditions these last relations become

(41) $\qquad\qquad y_1^2 + y_2^2 = 1, \qquad k^2 y_1^2 + y_3^2 = 1$

or

(41') $\qquad\qquad \text{sn}^2 x + \text{cn}^2 x = 1, \qquad k^2 \text{sn}^2 x + \text{dn}^2 x = 1$

From the first of these two fundamental equalities we deduce immediately, exactly as in the preceding section, that *the two functions* sn x *and* cn x

oscillate between -1 and $+1$, *and that when one of these two functions vanishes the other passes through its maximum value* $+1$ *or its minimum value* -1. From the second equation of (41') it follows, on the other hand, that $\mathrm{dn}^2 x$ oscillates between the narrower limits $1 - k^2$ and 1, and this implies that dn x never vanishes since we have supposed $k^2 < 1$. Since $\mathrm{dn}\, 0 = 1$ (which excludes the possibility that dn x can be negative) on writing

$$(42) \qquad +\sqrt{(1-k^2)} = k'$$

we deduce immediately that *the function* dn x *oscillates between the positive values* k' *and* 1, and attains these values when $\mathrm{sn}\, x = \pm 1$ or $\mathrm{sn}\, x = 0$ respectively.*

On introducing an angle ϕ (called the *amplitude* of the elliptic functions) such that

$$(43) \qquad \mathrm{sn}\, x = \sin \phi$$

and which, vanishing for $x = 0$, increases continuously as x increases, we deduce as a further corollary of (41') that

$$\mathrm{cn}\, x = \pm \cos \phi, \qquad \mathrm{dn}\, x = \pm \sqrt{(1 - k^2 \sin^2 \phi)}$$

and consequently that

$$(44) \qquad \mathrm{cn}\, x = \cos \phi, \qquad \mathrm{dn}\, x = \sqrt{(1 - k^2 \sin^2 \phi)}$$

since dn x is always positive, and the sign of cn x is determined by the equations, simultaneously valid,

$$\mathrm{sn}'\, x = \mathrm{cn}\, x \, \mathrm{dn}\, x, \qquad \mathrm{sn}'\, x = \cos \phi \frac{d\phi}{dx}$$

Since $d\phi/dx > 0$, cn x and $\cos \phi$ must always be of the same sign.**

The formulæ (43) and (44) explain the names (and also the symbols) usually attached to the three functions we are considering, namely *sine-amplitude*, *cosine-amplitude* and *delta-amplitude*. It is usual to denote by $\Delta\phi$ (delta ϕ) the function $\sqrt{(1 - k^2 \sin^2 \phi)}$.

We now ask whether the functions sn x and cn x can in fact vanish. As in the preceding case, the answer is obviously yes for sn x, which vanishes at least for $x = 0$—but what of cn x?

---

* The positive number k is called the *modulus* of the elliptic functions, and k' the *complementary modulus*.

** From these two formulæ it follows that

$$\frac{d\phi}{dx} = \mathrm{dn}\, x = \sqrt{(1 - k^2 \sin^2 \phi)}$$

It is convenient and in accordance with traditional usage to denote by K the first positive zero of cn x; we do not exclude the possibility that K may be $+\infty$ (this would be equivalent to saying that the function never vanishes for positive x) and we try to show that K is finite (provided $k^2 < 1$) by actually evaluating this number by a process similar to that employed in the preceding section to show that

$$\alpha = \tfrac{1}{2}A = \int_0^1 \frac{dx}{\sqrt{(1-x^2)}} = \frac{\pi}{2}$$

In view of (41), it follows from the first equation (38) that

(45) $$\frac{dx}{dy_1} = \frac{1}{y_1'} = \frac{1}{y_2 y_3} = \frac{1}{\sqrt{[(1-y_1^2)(1-k^2 y_1^2)]}}$$

Since as x increases from 0 to K the function cn x (i.e. y_2) decreases from 1 to 0, while the function sn x (i.e. y_1) increases from 0 to 1, we deduce that

(46) $$K = \int_0^1 \frac{dy_1}{\sqrt{[(1-y_1^2)(1-k^2 y_1^2)]}}$$

This formula shows that for $k^2 < 1$, K is in fact finite, for the integrand has no singularity in the interval of integration with the exception of an infinity of order $1/2$ at the point $y_1 = 1$. If on the other hand $k^2 = 1$, the integrand reduces to $(1 - y_1^2)^{-1}$ and K has a logarithmic infinity; this accords with the fact that the integral (46), which is obviously an increasing function of k^2, tends to infinity as $k^2 \to 1$.

At the point $x = K$ we evidently have

$$y_1(K) \equiv \operatorname{sn} K = 1,^* \qquad y_2(K) \equiv \operatorname{cn} K = 0, \qquad y_3(K) \equiv \operatorname{dn} K = k'$$

Analogy with the preceding case now suggests investigation of what results on changing the origin to the point $x = K$ by making the substitution

$$x = \xi + K$$

As it is desirable to keep the initial conditions unaltered we use a new transformation into unknown functions, less simple than that employed in the last section. Here the transformation is

$$y_1(x) = \frac{z_2(\xi)}{z_3(\xi)}, \qquad y_2(x) = -k'\frac{z_1(\xi)}{z_3(\xi)}, \qquad y_3(x) = \frac{k'}{z_3(\xi)}$$

---

* $y_1(K) = +1$ since $y_1(x)$ is an increasing function of x (this follows from the first equation (38)) in the interval $(0, K)$.

under which, as is easily verified, system (38) remains unchanged and may be written as

$$z'_1 = z_2 z_3, \qquad z'_2 = -z_1 z_3, \qquad z'_3 = -k^2 z_1 z_2$$

Further, the initial conditions remain unaltered as the transformation yields, inversely,

$$z_1(\xi) = -\frac{y_2(x)}{y_3(x)}, \qquad z_2(\xi) = k'\frac{y_1(x)}{y_3(x)}, \qquad z_3(\xi) = \frac{k'}{y_3(x)}$$

whence follow for $\xi = 0$, i.e. for $x = K$,

$$z_1(0) = 0, \qquad z_2(0) = 1, \qquad z_3(0) = 1$$

The theorem of uniqueness therefore implies that z_1, z_2, z_3 considered as functions of ξ must be in fact the functions y_1, y_2, y_3 of x; consequently we derive the important formulæ

(47) $$\operatorname{sn}(\xi+K) = \frac{\operatorname{cn}\xi}{\operatorname{dn}\xi}, \qquad \operatorname{cn}(\xi+K) = -k'\frac{\operatorname{sn}\xi}{\operatorname{dn}\xi}, \qquad \operatorname{dn}(\xi+K) = \frac{k'}{\operatorname{dn}\xi}$$

which replace (32) of the preceding section.

From the formulæ obtained above may be deduced several important results, particularly, as in the preceding section, that of the *periodicity* of the functions we are considering.

It follows immediately that

(48) $$\begin{cases} \operatorname{sn}(x+2K) = \dfrac{\operatorname{cn}(x+K)}{\operatorname{dn}(x+K)} = -\operatorname{sn} x \\[6pt] \operatorname{cn}(x+2K) = -k'\dfrac{\operatorname{sn}(x+K)}{\operatorname{dn}(x+K)} = -\operatorname{cn} x \\[6pt] \operatorname{dn}(x+2K) = \dfrac{k'}{\operatorname{dn}(x+K)} = \operatorname{dn} x \end{cases}$$

and successively that

$$\operatorname{sn}(x+4K) = -\operatorname{sn}(x+2K) = \operatorname{sn} x$$

$$\operatorname{cn}(x+4K) = -\operatorname{cn}(x+2K) = \operatorname{cn} x$$

Thus *the functions* $\operatorname{sn} x$ *and* $\operatorname{cn} x$ *are periodic with period* $4K$, *while the function* $\operatorname{dn} x$ *is periodic with period* $2K$.

On the other hand, it follows from formulæ (47) that

$$\operatorname{sn}(2K) = \frac{\operatorname{cn} K}{\operatorname{dn} K} = 0, \qquad \operatorname{cn}(3K) = -k'\frac{\operatorname{sn}(2K)}{\operatorname{dn}(2K)} = 0$$

$$\operatorname{sn}(4K) = \frac{\operatorname{cn}(3K)}{\operatorname{dn}(3K)} = 0, \qquad \ldots$$

Thus *the even multiples of K are zeros of* sn *x while the odd multiples are zeros of* cn *x*; further, these are *the only (real) zeros* of such functions, as is immediately seen by repetition of the same argument as that used in the analogous case in the preceding section. It further follows that *the even multiples of K give the maximum values of the function* dn *x while the odd multiples of K give the minimum values of the same function*.

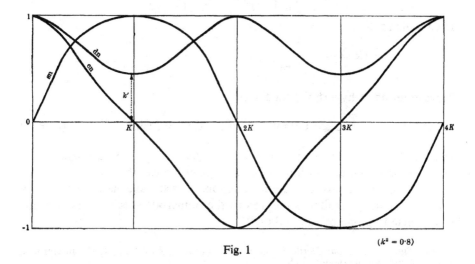

Fig. 1 $(k^2 = 0\cdot 8)$

The three functions behave, in general, as shown in figure 1 which illustrates the case in which $k^2 = 0\cdot 8$.

For elliptic functions also there are some addition theorems, similar to but more complicated than those for circular functions, and the derivation of these formulæ for elliptic functions by the methods used in the preceding section is made much more difficult by the fact that system (38), in contrast to system (26), is *non-linear*.

We note in conclusion that the elliptic functions may be considered as *inverse functions of integrals of algebraic functions*, which are in fact called *elliptic integrals*.

For example, from (45) it follows immediately that the function $\operatorname{sn} x \equiv y_1(x)$ is the inverse of the indefinite integral

$$(49) \qquad x = \int_0^{y_1} \frac{dy_1}{\sqrt{[(1-y_1^2)(1-k^2 y_1^2)]}}$$

From this formula it follows (as may be *read off directly* from the differential system) that *for $k^2 = 0$ the elliptic functions degenerate into circular functions*, i.e.

$$(50) \qquad \operatorname{sn} x = \sin x, \qquad \operatorname{cn} x = \cos x, \qquad \operatorname{dn} x = 1, \qquad (k^2 = 0)$$

If on the other hand $k^2 = 1$, (49) becomes

$$x = \int_0^{y_1} \frac{dy_1}{1-y_1^2} = \log\sqrt{\left(\frac{1+y_1}{1-y_1}\right)}$$

from which follow

$$\operatorname{sn} x = y_1 = \frac{e^x - e^{-x}}{e^x + e^{-x}}, \qquad \operatorname{cn} x = \operatorname{dn} x = \frac{2}{e^x + e^{-x}}$$

Thus in terms of hyperbolic functions

$$(50') \qquad \operatorname{sn} x = \tanh x, \qquad \operatorname{cn} x = \operatorname{dn} x = \operatorname{sech} x \qquad (k^2 = 1)$$

showing that *in the case $k^2 = 1$ the elliptic functions degenerate into hyperbolic functions* and are no longer periodic (K becomes *infinite*).

For a deeper study of elliptic functions which have here been merely touched upon, the reader is referred to the numerous special works on this topic, including one by the author.*

---

* F. TRICOMI, *Funzioni Ellittiche*, 2nd edition, Bologna, Zanichelli, 1951; in German, Leipzig Akad. Verslagsgesell., 1948.

II. The behaviour of the characteristics of a first-order equation

7. Preliminary considerations

It follows from the fundamental theorem of existence and uniqueness established in the preceding chapter, as a particular case for the first-order differential equation

$$\frac{dy}{dx} = f(x, y) \tag{1}$$

that through every point (x_0, y_0) of a domain D in which the function $f(x, y)$ is continuous and satisfies the Lipschitz condition in y there passes *exactly one characteristic* of the equation. (The *characteristics* of a differential equation are the curves which represent its integrals geometrically in the (x, y)-plane.) In general these curves cover the entire plane and do not intersect each other except in the possible *singular points* of (1), i.e. in the possible points in which the function ceases to be continuous or at least to satisfy the Lipschitz condition. For example, figure 2 illustrates the case of the equation

$$\frac{dy}{dx} = \frac{y}{2x} \tag{2}$$

(evidently possessing a singular point at the origin) whose characteristics are the parabolas

$$y^2 = cx$$

with common vertex the origin, as can be verified on integrating the equation by separating the variables.

In this chapter we propose to study mainly the types of singular points most frequently encountered in dealing with an equation of type (1), as familiarity with these singularities is of decisive importance in obtaining an idea of the general behaviour of the characteristics of such an equation.

Even the most simple considerations may be of great help in the discussion of these characteristics, as, for example, that the curve

(3) $$f(x, y) = 0$$

is obviously the locus of *stationary points* of the characteristics, i.e. the points at which the tangents are horizontal, and which, in general, are maximum or minimum points of the relative integral curves.

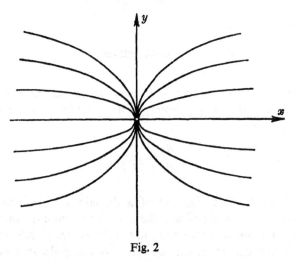

Fig. 2

It is also of some help to consider the locus of *points of inflection* of the characteristics, which since

$$\frac{d^2y}{dx^2} = f_x(x, y) + f_y(x, y)y'$$

is given by the equation

(4) $$f_x(x, y) + f_y(x, y) \cdot f(x, y) = 0$$

This curve, in general, divides the plane into two regions, in one of which the concavity of the characteristics is *upward*, in the other the concavity *downward*.

For example, in the case of the linear equation of the first order (with no finite singular points)

(5) $$y' = 1 + xy$$

the general integral is given by*

(6) $$y = e^{x^2/2}\left(c + \int_0^x e^{-x^2/2}\,dx\right)$$

* See, for example, F. Tricomi (79), part II, chapter IX, § 9.

As the indefinite integral* involved is not expressible in terms of elementary transcendental functions, this formula is not of much help in discussion of the behaviour of the characteristics. It is however sufficient to draw the curves (3) and (4) which are respectively the rectangular hyperbola

$$xy = -1$$

and the cubic curve

$$y + x(xy+1) = 0$$

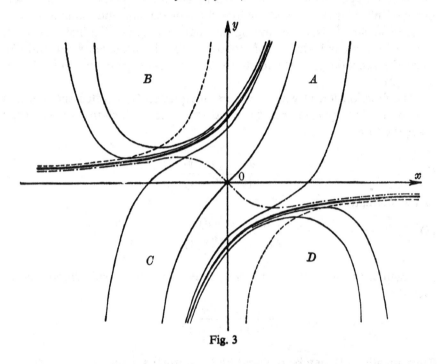

Fig. 3

(see figure 3 in which these two curves are drawn dotted) to see the general behaviour of the characteristics of (5). In the four regions A, B, C, D into which these two curves divide the (x, y)-plane, the characteristics must necessarily be (for increasing x)

> in A: ascending and concave upwards
> in B: descending and concave upwards
> in C: ascending and concave downwards
> in D: descending and concave downwards.

This indicates (as can be established rigorously by a more detailed argument) that there are two kinds of characteristics: (1) those of positive

---

* This is the *probability integral*. For further information see, for example, E. JAHNKE-F. EMDE (66).

gradient which lie entirely within the regions A and C and intersect the curve of inflection points and the x-axis but not the hyperbola of maxima and minima points; (2) those always concave upwards lying in the regions A and B, or always concave downwards lying in the regions C and D, which cut the hyperbola $xy = -1$ in exactly one point where, as can be verified, there is a *minimum* in the first case and a *maximum* in the second case.

This indicates that there are two special characteristics (indicated in the figure by heavy lines) having the x-axis as asymptote (as have also the cubic curve of inflection points and the hyperbola of maxima and minima points), these two characteristics separating the characteristics of the first kind from those of the second kind. This follows from the probability integral (6), since the function y in (6) tends to zero as $x \to \pm\infty$ if and only if $c = \mp\sqrt{(\pi/2)}$.

The function $f(x, y)$ which appears in equation (1) is often the quotient of two functions $Q(x, y)$ and $P(x, y)$, as in example (2), so that the equation takes the form

(7)
$$\frac{dy}{dx} = \frac{Q(x, y)}{P(x, y)}$$

or the alternative form

(8)
$$\frac{dx}{P(x, y)} = \frac{dy}{Q(x, y)}$$

It is helpful to introduce a new independent variable t with differential dt such that

$$\frac{dx}{P(x, y)} = \frac{dy}{Q(x, y)} = dt$$

Thus equation (7) may be replaced by the normal system

(9)
$$\begin{cases} \dfrac{dx}{dt} = P(x, y) \\ \dfrac{dy}{dt} = Q(x, y) \end{cases}$$

in which the variable t does not appear explicitly on the right-hand sides.

But is system (9) exactly equivalent to equation (7)?

Evidently there is equivalence provided that $P(x, y) \neq 0$; but if $P(x, y) = 0$, for example if $P(x, y)$ vanishes identically (with respect to y) for some particular value x_0 of x, system (9) is seen to be wider than equation (7) as

by associating with the constant $x = x_0$ a function $y = \phi(t)$ satisfying the equation with variables separable

$$\frac{dy}{dt} = Q(x_0, y)$$

we obtain a solution of the system (9) which only in a rather broad sense may be considered as corresponding to a solution of (7), since for $x = x_0$ this latter equation is meaningless.

Thus for the preceding equation (2), the corresponding system of the form (9)

(10) $$\frac{dx}{dt} = 2x, \quad \frac{dy}{dt} = y$$

possesses the solution $x = 0$ (i.e. the y-axis) which would be 'lost' were we to consider only the original form of the equation.

To the question raised above the reply is therefore:—
System (9) is not exactly equivalent to equation (7) but is wider than (7) in the sense that while every solution of (7) is a solution of (9) there may exist solutions of (9) for which (7) cannot be said to be satisfied, for when $P(x, y)$ vanishes the right-hand side of (7) becomes meaningless.

For this reason it is preferable, as a rule, to consider system (9) rather than equation (7) and *we shall therefore consider as characteristics of (7) (in the proper sense or in the wide sense) all characteristics of (9)*, thus *not excluding as possible solutions lines parallel to the y-axis.*

We note finally that in the case with which we are mainly concerned in this chapter, that in which the functions $P(x, y)$ and $Q(x, y)$ vanish simultaneously in a singular point (x_0, y_0), system (9) possesses the *constant solution*

(11) $$x = x_0, \quad y = y_0$$

to which there corresponds no real characteristic of (7), since corresponding to (11) there is no curve in the (x, y)-plane but merely a single point. However, it may happen, as in the case illustrated in figure 2, that there are other characteristics, possibly even an infinity of characteristics, through (x_0, y_0) —this is not in contradiction with the general theorem of existence and uniqueness (which is in general applicable to (9) whether P and Q vanish or not) because, as we shall see, at the corresponding point (x_0, y_0) the corresponding value of t on the characteristics is not finite but is infinite.

Systems of type (9) are of great importance in *non-linear mechanics*. A mechanical system with one degree of freedom x subjected to a force which does not depend explicitly on the time t, is described by a differential equation of the form

$$\frac{d^2x}{dt^2} = f\left(x, \frac{dx}{dt}\right)$$

and, putting $dx/dt = y$, the corresponding system in the (x, y)-phase-space may be written as

$$\frac{dx}{dt} = y, \qquad \frac{dy}{dt} = f(x, y)$$

a system of type (9).

8. Examples of equations with singular points

In equation (2), to which figure 2 refers, the singular point at the origin is called a *node*, since through it there pass an infinity of characteristics of the equation; for this particular equation, in fact, *all* its characteristics pass through this singular point. This is also the case for the more general equation

(12) $$y' = \alpha \frac{y}{x}$$

where α is any positive constant, as the integral of (12) is

$$\log|y| = \alpha \log|x| + \text{constant}$$

giving

(13) $$y = c|x|^\alpha$$

If $\alpha < 1$, the phase trajectories about the origin will behave similarly to those of figure 2, since from (13)

$$y' = \pm c\alpha |x|^{\alpha - 1}$$

and therefore if $c \neq 0$, as in (2),

$$\lim_{x \to 0} y' = \infty$$

implying that all the characteristics (with the exception of one characteristic, viz. the line $y = 0$) have the same tangent $x = 0$ at the origin. If on the other hand $\alpha > 1$, on interchanging x and y in (12) and transforming α into $1/\alpha$, the graphical representation of the characteristics about the origin is seen to be that obtained by rotating figure 2 through ninety degrees.

§ 8] Examples of equations with singular points

In both cases, there are exactly two possible directions of the tangents to the characteristics at the node, those along the x- and y-axes; these are called the *privileged directions* at that singular point.

Finally if $\alpha = 1$ there arises the exceptional case of the *star-shaped node* in which each characteristic passing through the node (i.e. each straight line through the origin) has its own tangent at the singular point; there are therefore no privileged directions in this case.

So far we have been concerned with $\alpha > 0$. If, instead, $\alpha < 0$* these results are radically changed, since then, as follows from (13), if $c \neq 0$,

$$\lim_{x \to 0} y = \infty$$

and in this case the characteristics, with the exception of *two*, do not pass through the origin—these exceptional characteristics being the line $y = 0$ (in the narrow sense of characteristic) and the line $x = 0$ (in the broader

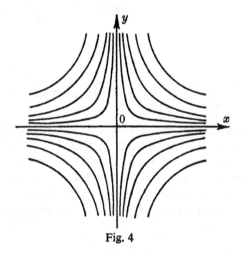

Fig. 4

meaning of the word). These results are illustrated in figure 4, in which α is taken as -1, and the characteristics are therefore the rectangular hyperbolas $xy = c$. As these somewhat resemble the contour lines on a topographic map in the vicinity of a saddle-point (French, *col*) this type of singular point is called a *col*.** In this case also there are two privileged directions, those of the two coordinate axes, in the sense that these are the tangential directions of the only two characteristics (here straight lines) which pass through the singular point.

* The case $\alpha = 0$, in which $y =$ constant, is of little interest.
** See fig 12., p. 69.

Another important equation with a *node* at the origin is

(14) $$\frac{dy}{dx} = \frac{x+y}{x}$$

On putting $y = xz$ this equation becomes

$$x\frac{dz}{dx} + z = 1 + z$$

i.e.

$$\frac{dz}{dx} = \frac{1}{x}$$

whence

$$z = \log|x| + \text{constant}$$

and therefore

(15) $$y = x(\log|x| + c)$$

In this case

$$\lim_{x \to 0} y = 0$$

and

$$\lim_{x \to 0} y' = \lim_{x \to 0} (1 + c + \log|x|) = -\infty$$

Hence all the characteristics pass through the origin and all (with no exceptions) have the y-axis as common tangent there; the y-axis is therefore the only privileged direction. Figure 5 in which the coordinate axes are denoted by ξ and η in place of x and y (as is also done in the two following figures) shows the behaviour of the characteristics in this case.

A very different behaviour is exhibited by the trajectories corresponding to the equation

(16) $$\frac{dy}{dx} = \frac{x+ay}{ax-y}$$

where a denotes a constant, which we meantime assume non-zero.

On changing to polar coordinates by the substitutions

$$x = \rho\cos\theta, \quad y = \rho\sin\theta$$

which imply

$$\rho^2 = x^2 + y^2, \quad \theta = \tan^{-1}\frac{y}{x}$$

$$\rho\frac{d\rho}{dx} = x + y\frac{dy}{dx}, \quad \rho^2\frac{d\theta}{dx} = x\frac{dy}{dx} - y$$

Examples of equations with singular points

and consequently

$$\frac{d\rho}{d\theta} = \rho \frac{x+yy'}{xy'-y}$$

we obtain in this case the very simple equation

$$\frac{d\rho}{d\theta} = a\rho$$

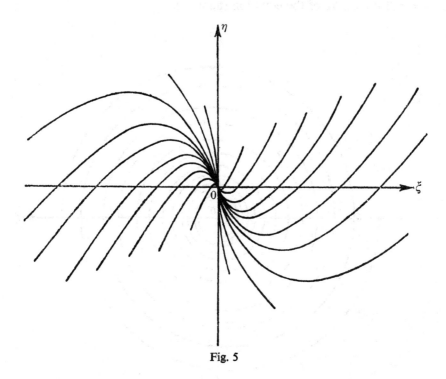

Fig. 5

whence, by separating the variables,

$$\log \rho = a\theta + \text{constant}$$

i.e.

(17) $$\rho = ce^{a\theta}$$

These characteristics are *logarithmic spirals* winding around the origin which is attained as limit as $\theta \to -\infty$ or $\theta \to +\infty$ according as $a > 0$ or $a < 0$ (figure 6). A singular point of this nature is called a *focus*. (The equivalent German term is *Strudelpunkt*, from *Strudel*, meaning a vortex.)

In the particular case $a = 0$ we derive immediately from (17), or even directly from (16) which in this case is

(18) $$\frac{dy}{dx} = -\frac{x}{y}$$

that $$xdx + ydy = \tfrac{1}{2}d(x^2 + y^2) = 0$$

Hence the characteristics are *circles* with centre the origin, which is for this reason called a *centre* of the equation (figure 7).

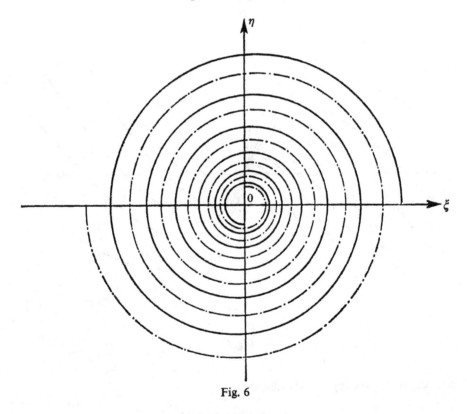

Fig. 6

Finally we consider an example of singular points of a completely different kind from any of the preceding—that of points at which the function $f(x, y)$ on the right-hand side of (7) is continuous but no longer satisfies the Lipschitz condition in the variable y.

For this purpose, consider the equation

(19) $$y' = +\sqrt{|y|}$$

i.e. let $$f(x, y) = +\sqrt{|y|}$$

§ 8] Examples of equations with singular points

This function $f(x, y)$ is continuous about all points on the x-axis but does not satisfy the Lipschitz condition at these points, since

$$\frac{f(x, y) - f(x, 0)}{y} = \frac{\sqrt{|y|}}{y} = \pm \frac{1}{\sqrt{|y|}}$$

which is unbounded as $y \to 0$.

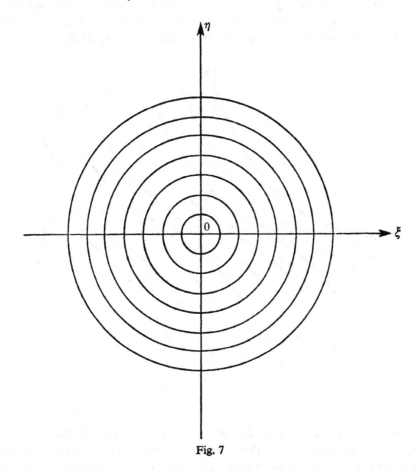

Fig. 7

Integration of (19), by separating the variables, gives

$$x + c = \int \frac{dy}{\sqrt{|y|}} = \begin{cases} +2\sqrt{y} & \text{(if } y \geqq 0) \\ -2\sqrt{(-y)} & \text{(if } y \leqq 0) \end{cases}$$

whence

$$y = \begin{cases} \tfrac{1}{4}(x+c)^2 \\ -\tfrac{1}{4}(x+c)^2 \end{cases}$$

or on replacing c by $-x_0$

(20) $$y = \begin{cases} \tfrac{1}{4}(x-x_0)^2 & (x \geqq x_0) \\ -\tfrac{1}{4}(x-x_0)^2 & (x \leqq x_0) \end{cases}$$

Thus a characteristic C_0 of (19) passing through the point $P_0(x = x_0, y = 0)$ (from which all the other characteristics of the equation, save one,* may be deduced by translations parallel to the y-axis) is made up of two half-parabolas both tangential to the x-axis at the common vertex P_0 (see figure 8). The x-axis also is obviously a characteristic of the equation.

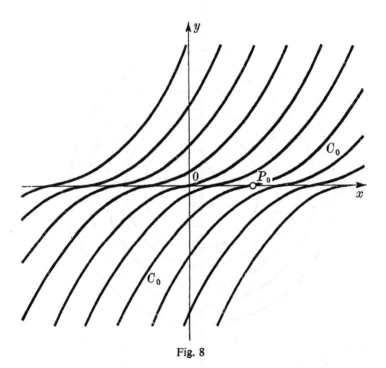

Fig. 8

This is sufficient to show that in this case the points of the x-axis are singular, in the sense that while the existence theorem of Chapter I remains valid at these points the uniqueness theorem fails—for we may consider as issuing from the point P_0, in addition to the characteristic C_0 mentioned above and the x-axis, the infinity of characteristics made up of arbitrary segments (x_1, x_2) of the x-axis (each containing within it the point x_0) and the two half-parabolas

(21) $y = -\tfrac{1}{4}(x-x_1)^2$, (for $x \leqq x_1$); $y = \tfrac{1}{4}(x-x_2)^2$, (for $x \geqq x_2$)

---

* The x-axis; see the following argument.

9. Study of the abridged equation

The different patterns exhibited in the preceding section, although fairly simple, make apparent the usefulness of some criterion for the behaviour of the characteristics of an equation of type (7) in the vicinity of a point (x_0, y_0) at which $P(x, y)$ and $Q(x, y)$ vanish simultaneously, and consequently the existence and uniqueness theorem of Chapter I fails entirely.*

Work on these singular points—begun by H. Poincaré at the end of the last century and continued by I. Bendixson in a classic memoir of 1901, and more recently by several well-known contemporary mathematicians—was initially based on rather restrictive hypotheses on the functions P and Q which were assumed *analytic* (i.e. developable in power series) or simply *polynomials*. Gradually these conditions have been weakened until in the last few years they have been reduced in some cases to necessary conditions and in others to conditions only a little more restrictive. But this has been achieved only by complicated and difficult arguments.

We shall pursue a middle path by imposing on the functions $P(x, y)$ and $Q(x, y)$ conditions which while being weakly restrictive will yet allow a treatment both simple and rigorous. It is convenient in the first instance to suppose that the functions P and Q are *linear*, and that the singular point is the origin of coordinates; in place of equation (7) we therefore consider the *abridged equation*

$$(22) \qquad \frac{dy}{dx} = \frac{Cx + Dy}{Ax + By}$$

where

$$(23) \qquad P_x(0, 0) = A, \quad P_y(0, 0) = B, \qquad Q_x(0, 0) = C, \quad Q_y(0, 0) = D$$

In view of the comment at the end of § 7 we shall consider in this section the linear system

$$(24) \qquad \frac{dx}{dt} = Ax + By, \qquad \frac{dy}{dt} = Cx + Dy$$

where we suppose

$$(25) \qquad AD - BC \neq 0$$

---

* If only the denominator function $P(x, y)$ vanishes an interchange of x and y reduces the equation to a form in which no singularity appears and to which therefore the results of Chapter I may be applied. The only point to be specially noted here is that the characteristic which passes through the point (x_0, y_0) will have a *vertical* tangent at that point.

We propose to investigate whether the system (24) can be reduced to the *canonical form*

(26) $$\frac{d\xi}{dt} = \lambda_1 \xi, \qquad \frac{d\eta}{dt} = \lambda_2 \eta$$

where λ_1 and λ_2 denote two suitable *real* constants, by means of a linear substitution of the type

(27) $$\begin{cases} \xi = \alpha x + \beta y \\ \eta = \gamma x + \delta y \end{cases} \qquad (\alpha\delta - \beta\gamma \neq 0)$$

If so, by comparing (26) with the single equation

(28) $$\frac{d\eta}{d\xi} = \frac{\lambda_2 \eta}{\lambda_1 \xi}$$

which has already been discussed in the preceding section the results will follow immediately.

By replacing ξ and η in (26) by the expressions (27) and using (24) we find

$$\alpha(Ax+By) + \beta(Cx+Dy) = \lambda_1(\alpha x + \beta y)$$
$$\gamma(Ax+By) + \delta(Cx+Dy) = \lambda_2(\gamma x + \delta y)$$

and on equating the coefficients of x and y in these identities we obtain

(29') $$\begin{cases} (A-\lambda_1)\alpha + C\beta = 0 \\ B\alpha + (D-\lambda_1)\beta = 0 \end{cases}$$
(29'') $$\begin{cases} (A-\lambda_2)\gamma + C\delta = 0 \\ B\gamma + (D-\lambda_2)\delta = 0 \end{cases}$$

In order that these linear equations, which are homogeneous in α, β and γ, δ, may be satisfied by values of α, β, γ, δ not all zero, it is necessary and sufficient that the determinant of the coefficients vanishes, i.e. that λ_1, λ_2 are the roots of the *characteristic equation*

(30) $$\begin{vmatrix} A-\lambda & C \\ B & D-\lambda \end{vmatrix} = 0$$

i.e.
$$\lambda^2 - (A+D)\lambda + AD - BC = 0$$

which implies

(31) $$\lambda = \tfrac{1}{2}[(A+D) \pm \sqrt{(A-D)^2 + 4BC}]$$

Provided therefore that

$$(A-D)^2 + 4BC > 0$$

Study of the abridged equation

the values of λ_1, λ_2 given by (31) are *real and distinct* and the reduction to the canonical form (26) may be carried out.*

When (30) is satisfied we need consider only the first equations in (29′) and (29″) (as the second equations are merely derived from the first) and we may write simply

$$(32) \qquad \alpha = C, \quad \beta = \lambda_1 - A, \quad \gamma = C, \quad \delta = \lambda_2 - A$$

Thus, for $C \neq 0$,** $\alpha\delta - \beta\gamma \neq 0$ since, from (32), $\alpha\delta - \beta\gamma = (\lambda_2 - \lambda_1)C$.

Employing the results established in the preceding paragraph for equations of the type (28) and noting that λ_2/λ_1 has the sign of $\lambda_1\lambda_2$, i.e. of $AD - BC$, we can therefore state:

If the discriminant $(A - D)^2 + 4BC$ of the characteristic equation is positive, the origin is a NODE *or a* COL *of equation (22) according as $AD - BC$ is positive or negative.*

We now turn to the case in which the discriminant $(A - D)^2 + 4BC$ is *negative*, and therefore the two roots of the characteristic equation are *conjugate complex*, i.e.

$$(33) \qquad \lambda_1 = \mu + iv, \quad \lambda_2 = \mu - iv$$

where μ and v are two real quantities and $v \neq 0$.

Since the preceding calculations are valid also in this case, where now complex quantities are involved, (27) together with (32) and (33) yield

$$(34) \qquad \begin{cases} \xi = Cx + (\mu - A + iv)y \\ \eta = Cx + (\mu - A - iv)y \end{cases}$$

This suggests using, in place of (27), the *real* transformations

$$(35) \qquad \begin{cases} X = Cx + (\mu - A)y \\ Y = vy \end{cases}$$

which imply

$$(36) \qquad \xi = X + iY, \quad \eta = X - iY$$

This substitution changes the given system (24) into another which is perhaps less simple than (26) but which nevertheless yields as exhaustive results as those derived in the preceding paragraph. By virtue of (33) and

---

* Note that neither λ_1 nor λ_2 can be zero since $\lambda_1\lambda_2 = AD - BC \neq 0$.

** It can be assumed that $C \neq 0$ since if $C = 0$ then either $B = 0$, in which case system (24) is already in canonical form, or $B \neq 0$ in which case it is sufficient to interchange x and y (and therefore B and C) to arrange that the condition is satisfied.

(36), equations (26) now take the form

$$\begin{cases} \dfrac{dX}{dt} + i\dfrac{dY}{dt} = (\mu X - \nu Y) + i(\nu X + \mu Y) \\ \dfrac{dX}{dt} - i\dfrac{dY}{dt} = (\mu X - \nu Y) - i(\nu X + \mu Y) \end{cases}$$

and are therefore satisfied by simply putting

(37)
$$\begin{cases} \dfrac{dX}{dt} = \mu X - \nu Y \\ \dfrac{dY}{dt} = \nu X + \mu Y \end{cases}$$

In other words, in the case in which the discriminant of the characteristic equation is negative it is convenient to use, in place of the substitution (27) with coefficients (32), the linear substitution (35) which reduces the given system to the new canonical form (37), i.e. to the single equation

(38) $$\frac{dY}{dX} = \frac{X + aY}{aX - Y}$$

where $a = \mu/\nu$. But (38) is exactly equation (16) of the preceding section: thus

In the case in which the discriminant $(A - D)^2 + 4BC$ of the characteristic equation is negative, the origin is a FOCUS *of equation* (22), *provided that $A + D \neq 0$;* in the case in which $A + D = 0$ the origin is a* CENTRE.

It only remains to consider the case in which the discriminant is zero, i.e. in which

(39) $$(A - D)^2 + 4BC = 0$$

and therefore

(40) $$\lambda_1 = \lambda_2 = \tfrac{1}{2}(A + D)$$

Under these hypotheses and assuming in the first instance that at least one of the two quantities B and C (and we may always take C) is non-zero, for the value $\tfrac{1}{2}(A + D)$ of λ given by (40) we *cannot* determine a set of values of α, β, γ, δ (not all zero and for which $\alpha\delta - \beta\gamma \neq 0$) such that systems (29′) and (29″) are simultaneously satisfied. But we can obtain

---

* The condition $A + D = 0$ implies $\mu = 0$.

§ 9] Study of the abridged equation

values such that at least one of these two systems is satisfied, for example (29′). For this it is convenient to give α and β the values

(41) $$\alpha = C, \quad \beta = \lambda_1 - A = \tfrac{1}{2}(D-A)$$

which ensures that the first of the two differential equations in ξ and η takes the form (26), i.e.

(42) $$\frac{d\xi}{dt} = \lambda_1 \xi$$

As regards the second equation we try to reduce it to the form

(42′) $$\frac{d\eta}{dt} = \lambda_1(\xi + \eta)$$

by suitable choices of γ and δ. By a process similar to that used above we obtain the equalities

(43) $$\begin{cases} (A-\lambda_1)\gamma + C\delta = \lambda_1 \alpha \\ B\gamma + (D-\lambda_1)\delta = \lambda_1 \beta \end{cases}$$

which are in reality a single equality; for if $A \neq D$ and hence $\beta \neq 0$, the common value*

$$\frac{A-\lambda_1}{B} = \frac{C}{D-\lambda_1}$$

of the ratios of the coefficients of γ and δ on the left-hand sides equals the ratio of the right-hand sides as

$$\frac{\lambda_1 \alpha}{\lambda_1 \beta} = \frac{C}{\tfrac{1}{2}(D-A)} = \frac{C}{D-\lambda_1}$$

while if $A = D$ (and hence $B = \beta = 0$) the second equation (43) is satisfied identically.

It follows from this that it is possible to satisfy equations (43), for example by taking

$$\gamma = 0, \quad \delta = \frac{\lambda_1 \alpha}{C} = \tfrac{1}{2}(A+D)$$

We conclude therefore that the linear substitution

(44) $$\begin{cases} \xi = Cx - \tfrac{1}{2}(A-D)y \\ \eta = \tfrac{1}{2}(A+D)y \end{cases}$$

---

* Since λ_1 is the (double) root of equation (30) the determinant formed by the coefficients of (30) is zero.

with determinant $\frac{1}{2}(A + D)C$, which is certainly non-zero since

$$C \neq 0, \quad \tfrac{1}{4}(A+D)^2 = \lambda_1^2 = AD-BC \neq 0$$

reduces the differential system (24) to the canonical form (42)-(42'), i.e. to the single equation

(45) $$\frac{d\eta}{d\xi} = \frac{\xi+\eta}{\xi}$$

which is equation (14) of § 8. We thus conclude that *the origin is a* NODE *with unique tangent the η-axis.*

It remains finally to examine the subcase for which $B = C = 0$; in this case since evidently $A = D$ the equation has a form which we have already discussed, viz.

$$\frac{dy}{dx} = \frac{y}{x}$$

which is equation (12) of § 8 with $\alpha = 1$. We therefore have a *star-shaped* NODE.

Summarizing the results now obtained we therefore have:

The equation

$$\frac{dy}{dx} = \frac{Cx+Dy}{Ax+By}$$

with $AD - BC \neq 0$ has a singularity at the origin; the type of singularity is specified as follows:

$$\begin{cases} (A-D)^2+4BC>0 & \begin{cases} AD-BC>0 & \text{node} \\ AD-BC<0 & \text{col} \end{cases} \\ (A-D)^2+4BC<0 & \begin{cases} A+D \neq 0 & \text{focus} \\ A+D = 0 & \text{centre} \end{cases} \\ (A-D)^2+4BC = 0 & \text{node (a } \textit{star-shaped node} \text{ if and only if } A = D, \\ & B = C = 0). \end{cases}$$

The following points should be noted:

(1) To conclude that the singularity is a *col* it is sufficient to verify only that $AD - BC < 0$, since then

$$(A-D)^2+4BC = (A+D)^2-4(AD-BC)>0$$

(2) In the case of a *col*, as in that of a *node*, the gradients of the two privileged directions are the roots of the quadratic equation

$$B\mu^2 + (A-D)\mu - C = 0$$

obtained from (22) on writing

$$\frac{dy}{dx} = \frac{y}{x} = \mu$$

(3) In the case so far excluded, in which

$$AD - BC = 0$$

i.e. $C/A = D/B$, the abridged equation (22) reduces simply to

$$\frac{dy}{dx} = k$$

where k is the common value of these two ratios; hence its characteristics are the lines of gradient k and there is no longer a singularity at the origin. We shall see later (p. 71) how very different may be the behaviour in the case of a non-linear equation.

10. Some theorems of a general character

The principal aim of the present chapter is to show how the results obtained in the preceding section for the particular case dealt with there may be applied, *in general* unchanged, to a differential system of the form

(46) $$\frac{dx}{dt} = P(x, y), \quad \frac{dy}{dt} = Q(x, y)$$

where the functions $P(x, y)$ and $Q(x, y)$ are suitably restricted while the equations (23) of the preceding section still remain valid, i.e. to show how, in general, the nature of the singularity depends only on the terms of first degree in the functions $P(x, y)$ and $Q(x, y)$.

Systems of the form (46) in which the independent variable t does not appear on the right-hand side are usually called *autonomous systems*.

It is convenient to establish first some general properties of the characteristics of a system of the form (46); in addition to their intrinsic interest these properties will be of help in obtaining the final result.

The first of these properties is concerned with the *continuability* of a characteristic c, with parametric equations

(47) $$x = \phi(t), \qquad y = \psi(t)$$

issuing from a point $P_0(x_0, y_0)$ in a domain D within which P and Q satisfy conditions under which the fundamental theorem of Chapter I is valid,* i.e. they are *continuous and satisfy the Lipschitz condition* (with respect to both x and y). Under such hypotheses the fundamental theorem establishes that, if t_0 be the value of the parameter t corresponding to the point P_0, the functions $\phi(t)$ and $\psi(t)$ can be defined for $t > t_0$ at least in a certain interval (t_0, t_1) for whose length a lower bound is given explicitly in Chapter I; and similarly for $t < t_0$. It follows from this, that if the point P_1 with coordinates

$$x_1 = \phi(t_1), \qquad y_1 = \psi(t_1)$$

also lies within the domain D, the fundamental theorem may be applied a second time about the point (x_1, y_1), thus *extending* the functions $\phi(t)$ and $\psi(t)$ into a new interval (t_1, t_2) consecutive to (t_0, t_1), and so on. Thus is generated an increasing sequence $t_0, t_1, t_2, \ldots$, of values of t, which is either *divergent*, in which case the corresponding characteristic may be continued indefinitely, or the sequence tends to a certain finite limit T (the *escape time*).

In the latter case, the point $\bar{P}$ with coordinates $(\phi(T), \psi(T))$ must be a frontier point of the domain D.

If this were not so, we could again apply the fundamental theorem about the point $\bar{P}$ and deduce that the functions $\phi(t)$ and $\psi(t)$ may be continued to the right of $t = T$, contrary to the hypothesis that T is the limit point, and thus the upper bound, of the increasing sequence $t_0, t_1, t_2, \ldots$ It follows therefore that

*Every semi-characteristic** issuing from a point P_0 in the domain D may be continued until the parameter t takes the value $\pm \infty$, or until the frontier of the domain D is reached.*

Of these two possibilities the more interesting is evidently the first since in the latter case it is clear that nothing can be said about the characteristic c after it leaves the domain D within which the functions $P(x, y)$ and $Q(x, y)$ satisfy the fundamental hypotheses, or perhaps within which alone P and Q are defined. We ask therefore what happens to the point P with coordinates

---

* We do not exclude the possibility that the domain D may contain singular points at which $P(x, y)$ and $Q(x, y)$ vanish simultaneously.

** When, as in this case, it is useful to distinguish the two parts into which a characteristic issuing from some point P_0 is cut by the point P_0 itself each part is called a *semi-characteristic*.

$x = \phi(t)$, $y = \psi(t)$ when $t \to +\infty$ or to $-\infty$, now supposing that this point remains within the domain D. We shall show that *if $\phi(t)$ and $\psi(t)$ tend to finite limits a and b as $t \to +\infty$ (or as $t \to -\infty$), i.e.*

$$\lim_{t \to +\infty} \phi(t) = a, \qquad \lim_{t \to +\infty} \psi(t) = b$$

the point (a, b) must be a singular point of the system (46), *i.e.*

$$P(a, b) = Q(a, b) = 0$$

Otherwise, if at least one of the functions P and Q does not vanish at the point (a, b), for example if P does not vanish, then

$$|P(a, b)| = \eta > 0$$

and we can find a t^* such that for all $t > t^*$

$$|P(x, y)| = \left|\frac{d\phi}{dt}\right| > \tfrac{1}{2}\eta;$$

this implies that, given t_1 and t_2 any two values of t greater than t^*, and t' a suitable value between t_1 and t_2, then

$$|\phi(t_2) - \phi(t_1)| = |(t_2 - t_1)\phi'(t')| > \tfrac{1}{2}\eta|t_2 - t_1|$$

Thus if $|t_2 - t_1| \geq 1$

$$|\phi(t_2) - \phi(t_1)| > \tfrac{1}{2}\eta$$

in contradiction to the hypothesis that $\phi(t)$ tends to a definite limit as $t \to +\infty$.

Conversely, *if as $t \to t_0$ the characteristic c ends in a singular point (x_0, y_0) of D, i.e. if*

$$\lim_{t \to t_0} \phi(t) = x_0, \qquad \lim_{t \to t_0} \psi(t) = y_0; \qquad P(x_0, y_0) = Q(x_0, y_0) = 0$$

then $t_0 = \pm \infty$; for if, on the contrary, t_0 were finite, the uniqueness theorem of Chapter I would imply that (46) has no solution other than the trivial solution

$$x = x_0 = \text{constant}, \qquad y = y_0 = \text{constant}$$

to which corresponds no actual characteristic satisfying the initial conditions

$$x = x_0, \qquad y = y_0 \qquad\qquad \text{(for } t = t_0\text{)}$$

or the equivalent* conditions

$$\lim_{t \to t_0} x = x_0, \quad \lim_{t \to t_0} y = y_0$$

The preceding remark applies equally in the case in which the characteristic ends in (x_0, y_0) with a definite tangent, as happens, for example, for *nodes* and *cols*, and in the case in which it ends in a spiral as happens for *foci*. But are there any other possibilities? In answer to this question the following theorem due, as are many others in this chapter, to Bendixson establishes that under fairly weak conditions on the functions P and Q all other possibilities can be ruled out. Bendixson's original proof demanded that the functions P and Q be analytic; we now give a new proof of the result under the following weaker conditions:

*If the functions $P(x, y)$ and $Q(x, y)$, in addition to satisfying the conditions of continuity stated earlier, vanish at the origin** and are such that (uniformly)*

$$P(x, y) = H(x, y) + F(x, y), \quad Q(x, y) = K(x, y) + G(x, y)$$

where $H(x, y)$ and $K(x, y)$ denote two homogeneous polynomials OF THE SAME DEGREE $m \geq 1$ *without real common factors,**** and $F(x, y)$ and $G(x, y)$ are two functions such that*

(48) $$\lim_{\rho \to 0} \frac{F(x, y)}{\rho^m} = 0, \quad \lim_{\rho \to 0} \frac{G(x, y)}{\rho^m} = 0 \quad (\rho = \sqrt{[x^2 + y^2]})$$

and if, in addition, the homogeneous polynomial of degree $m + 1$

(49) $$xK(x, y) - yH(x, y) = M(x, y)$$

*is not identically zero, every characteristic of the system (46) which approaches the origin indefinitely**** either ends there in a spiral, or ends there with a well-defined tangent making with the x-axis an angle θ^* which satisfies the equation (immediately reducible to an algebraic equation of degree $m + 1$)*

(50) $$M(\cos \theta^*, \sin \theta^*) = 0$$

for on changing to polar coordinates

$$x = \rho \cos \theta, \quad y = \rho \sin \theta$$

---

* Under the hypotheses of Chapter I the integrals of a normal differential system are continuous functions of the initial values (p. 11).

** We again suppose, as in § 9, that the singular point is translated to the origin.

*** We suppose that the two polynomials do not contain a common divisor of the form $ax + by$, where a and b are real. In other words, neither the two algebraic equations $H(x, 1) = 0$, $K(x, 1) = 0$ nor the pair $H(1, y) = 0$, $K(1, y) = 0$ possess any real common root.

**** This means that on the characteristic, $\lim \rho = 0$ either as $t \to +\infty$ or $t \to -\infty$.

which implies
$$\rho^2 = x^2+y^2, \qquad \theta = \tan^{-1} y/x$$
and

(51) $$\rho\frac{d\rho}{dt} = x\frac{dx}{dt}+y\frac{dy}{dt}, \qquad \rho^2\frac{d\theta}{dt} = x\frac{dy}{dt}-y\frac{dx}{dt}$$

the system (46) becomes
$$\begin{cases} \rho\dfrac{d\rho}{dt} = xH(x, y)+yK(x, y)+xF(x, y)+yG(x, y) \\ \rho^2\dfrac{d\theta}{dt} = xK(x, y)-yH(x, y)+xG(x, y)-yF(x, y) \end{cases}$$

This may be put into the form

(52) $$\frac{1}{\rho^m}\frac{d\rho}{dt} = \Phi(\theta)+E_1(\rho, \theta), \qquad \frac{1}{\rho^{m-1}}\frac{d\theta}{dt} = \Psi(\theta)+E_2(\rho, \theta)$$

where

(53) $$\begin{cases} \Phi(\theta) = H(\cos\theta, \sin\theta)\cos\theta+K(\cos\theta, \sin\theta)\sin\theta \\ \Psi(\theta) = K(\cos\theta, \sin\theta)\cos\theta-H(\cos\theta, \sin\theta)\sin\theta = M(\cos\theta, \sin\theta) \end{cases}$$

and where $E_1(\rho, \theta)$ and $E_2(\rho, \theta)$ denote two functions which tend uniformly to zero as $\rho \to 0$, viz.

(54) $$\begin{cases} E_1(\rho, \theta) = [F(\rho\cos\theta, \rho\sin\theta)\cos\theta+G(\rho\cos\theta, \rho\sin\theta)\sin\theta]\rho^{-m} \\ E_2(\rho, \theta) = [G(\rho\cos\theta, \rho\sin\theta)\cos\theta-F(\rho\cos\theta, \rho\sin\theta)\sin\theta]\rho^{-m} \end{cases}$$

Now let us suppose that a certain characteristic c, represented by the parametric equations $x = \phi(t)$, $y = \psi(t)$, *ends in the origin as $t \to +\infty$*, i.e. that
$$\lim_{t \to +\infty} \phi(t) = \lim_{t \to +\infty} \psi(t) = 0$$

This implies that $\rho \to 0$ as $t \to +\infty$, i.e. that given a positive number ε however small, we can always find a t^* such that $\rho < \varepsilon$ for all $t > t^*$. In other words, by considering θ and ρ as Cartesian coordinates in an auxiliary plane, we find that for $t > t^*$ the characteristic* will always remain within the strip determined by the θ-axis and the line $\rho = \varepsilon$ parallel to it (figure 9).

What happens to θ as $t \to +\infty$?

---

* It would be more correct to say 'the image of the characteristic in the auxiliary plane'.

There are three *a priori* possibilities:

(1) θ tends to $\pm\infty$ as $t \to +\infty$.
(2) θ tends to a finite limit θ^*.
(3) θ gives rise to two or more limit points $\theta_1^*, \theta_2^*, \ldots$.*

We shall first show that the third possibility cannot in fact occur.

Indeed if the third hypothesis were true then since the function $\Psi(\theta) = M(\cos\theta, \sin\theta)$ *is not identically zero* this function must vanish for at most $2m+2$ real and distinct values of θ lying between 0 and 2π and for all other values of θ obtained from these by adding or subtracting integral multiples of 2π. These values form a *discrete* set (i.e. without finite points of accumulation) of points on the θ-axis. Thus given θ_1^* any one of the limiting values

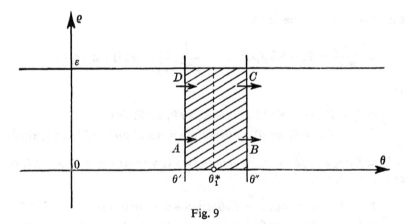

Fig. 9

of θ as $t \to +\infty$ we can find in its vicinity as many pairs θ', θ'' of values of θ as we please such that $\Psi(\theta') \neq 0$, $\Psi(\theta'') \neq 0$ and in addition θ' and θ'' differ from θ_1^*, by as small quantities as we please, one being less than the other being greater than θ_1^*.

It follows from this—since when the positive number ε is sufficiently small $d\theta/dt$ has the sign of $\Psi(\theta)$, as the second term of the right-hand side of the second equation in (52) tends uniformly to zero as $\rho \to 0$—that for increasing t the characteristics can cross only in one clearly defined direction the sides AD and BC of the rectangle defined in the strip $0 \leq \rho \leq \varepsilon$ by the parallels to the ρ-axis through the points $\theta = \theta'$ and $\theta = \theta''$ on the θ-axis; for example, the characteristics can cross these sides only in the direction indicated by the arrows in figure 9 when $\Psi(\theta') > 0$ and $\Psi(\theta'') > 0$.

---

* As $\theta \equiv \theta(t)$ is a continuous variable, to say that θ^* is a *limit point* (as $t \to \infty$) means the following: given any positive number ε as small as we please and any value t_0 of t we can always find *some* value t_1 of t to the right of t_0 such that $|\theta(t_1) - \theta^*| < \varepsilon$.

But this shows that a characteristic arising within the rectangle ABCD or penetrating into it, on once leaving this rectangle (if this is possible) can never re-enter the rectangle, in contradiction to the assumption that θ_1^* is a limit point of θ; therefore in order that a characteristic should give rise to a limit point θ_1^* it must always remain within the rectangle ABCD from some $\bar{t} \geq t^*$ onwards, i.e. it must be such that $\theta' \leq \theta \leq \theta''$ for $t > \bar{t}$. Thus, as θ' and θ'' may be chosen as close to each other and to θ_1^* as we wish, it follows that

$$\lim_{t \to +\infty} \theta = \theta_1^*$$

contrary to the hypothesis that θ_1^* is *one* of the limit points of θ as $t \to +\infty$. There can therefore be *only one* limit point of θ, finite or infinite.

This proves that when $t \to \infty$ any characteristic which approaches the origin indefinitely will end there either in the form of a spiral (the case in which $\lim \theta = \pm \infty$) or with a well-defined tangent (the case in which $\lim \theta = \theta^*$).

It still remains to show that the equation (50), which is actually an algebraic equation, must be satisfied, i.e. that

(55) $$\Psi(\theta^*) = 0$$

To show this we observe that as $t \to +\infty$ and $\rho \to 0$

$$\lim_{\rho \to 0} \frac{dy}{dx} = \lim_{\rho \to 0} \frac{K(x, y) + G(x, y)}{H(x, y) + F(x, y)} = \lim_{\rho \to 0} \frac{\rho^m K(\cos \theta, \sin \theta) + G(\rho \cos \theta, \rho \sin \theta)}{\rho^m H(\cos \theta, \sin \theta) + F(\rho \cos \theta, \rho \sin \theta)}$$

$$= \lim_{\rho \to 0} \frac{K(\cos \theta, \sin \theta) + \rho^{-m} G(\rho \cos \theta, \rho \sin \theta)}{H(\cos \theta, \sin \theta) + \rho^{-m} F(\rho \cos \theta, \rho \sin \theta)} = \frac{K(\cos \theta^*, \sin \theta^*)}{H(\cos \theta^*, \sin \theta^*)}$$

with the proviso that, if $H(\cos \theta^*, \sin \theta^*) = 0$, we must consider in place of this ratio the inverted ratio derived by interchanging x and y*. By l'Hospital's rule, we have

$$\tan \theta^* = \lim_{\rho \to 0} \frac{y}{x} = \lim_{\rho \to 0} \frac{dy}{dx}$$

and therefore

(55') $$\frac{K(\cos \theta^*, \sin \theta^*)}{H(\cos \theta^*, \sin \theta^*)} = \tan \theta^*$$

which is only (50) (or (55)) written in different form.

---

* If $H(\cos \theta^*, \sin \theta^*) = 0$ then $K(\cos \theta^*, \sin \theta^*) \neq 0$, for otherwise the polynomials $H(x, y)$ and $K(x, y)$ would possess the real common factor $x \sin \theta^* - y \cos \theta^*$ in contradiction of the initial hypothesis.

From this important result there follow various corollaries which allow us in several cases to settle without other consideration the behaviour of the characteristics which indefinitely approach a singular point.

COROLLARY I.—*If the hypotheses of Bendixson's theorem are satisfied and if equation (50), or (which is the same thing) the equation $\Psi(\theta) = 0$ possesses no real roots, every characteristic approaching the origin indefinitely must wind round it in the form of a spiral.*

Under these conditions the first of the three hypotheses made in the proof of the preceding theorem must hold, since (55) is impossible.

COROLLARY II.—*If the hypotheses of Bendixson's theorem are satisfied and if there is one characteristic which ends in the origin with a definite tangent, every characteristic approaching the origin indefinitely must behave similarly.*

There can be no characteristics of spiral form since any such must cut (infinitely many times) the special characteristic mentioned in points distinct from the singular point, and this is impossible by the fundamental theorem of Chapter I.

COROLLARY III.—*If under the same hypotheses as before the function $\Psi(\theta)$ is of varying sign, no characteristic can wind round the origin in the form of a spiral, and those characteristics which approach the origin indefinitely end there with a definite tangent.*

Under the above hypotheses we can find two angles θ_1 and θ_2 between 0 and 2π such that $\Psi(\theta_1) > 0$ and $\Psi(\theta_2) < 0$. Consequently—since the preceding function $E_2(\rho, \theta)$ tends to zero as $\rho \to 0$—if we choose a ρ_0 such that for $\rho \leqq \rho_0$

$$|E_2(\rho, \theta_1)| < \Psi(\theta_1), \qquad |E_2(\rho, \theta_2)| < -\Psi(\theta_2)$$

the second equation in (52) then implies that on the segments of the two radii $\theta = \theta_1$ and $\theta = \theta_2$ within the circle $\rho = \rho_0$ we have respectively $d\theta/dt > 0$, $d\theta/dt < 0$; hence, for increasing t, any characteristic can cross the first radius only in the direction of increasing θ (or the contrary for decreasing t). But under such conditions, if t is already sufficiently large in absolute value so that $\rho < \rho_0$, a characteristic once having entered into one of the two sectors into which the chosen radii divide the circle $\rho \leqq \rho_0$ (i.e. into that sector which the characteristic may enter) cannot again leave this sector; hence such a characteristic cannot wind round the origin in the form of a spiral, and it must therefore end in the origin with a definite tangent, which will be a radius $\theta = \theta^*$ for which $\Psi(\theta^*) = 0$.

We note finally that if the condition of Bendixson's theorem (viz. that the polynomial $M(x, y) = xK(x, y) - yH(x, y)$ is not identically zero) is not satisfied, i.e. if $\Psi(\theta) \equiv 0$, then, in general, characteristics issue from the origin in all possible directions, as happens for equation (12) of §8 for the case $\alpha = 1$ (*a star-shaped node*).

In a following section, we shall briefly deal with such a case when $m = 1$. If, instead, $m \geq 2$ we refer the reader to the classic memoir of Bendixson[*] to which reference has already been made, or to the more recent work of Lonn[**] to which we shall refer frequently in following sections.

11. The Poincaré index

It is now intended, as has already been suggested, to narrow the conditions imposed on the right-hand sides of the differential equations under consideration to obtain simple and neat results.

For this purpose we add to the hypotheses of the Bendixson theorem only the condition $m = 1$, i.e. we now suppose that the functions $P(x, y)$ and $Q(x, y)$ on the right-hand sides of equations (46) satisfy the following conditions:

(1) *The functions $P(x, y)$ and $Q(x, y)$ are continuous in some neighbourhood of the origin within which they also satisfy the Lipschitz condition either with respect to x or to y.*

(2) *There exist four constants A, B, C, D, not all zero, such that*

(56) $\qquad P(x, y) = Ax + By + F(x, y), \qquad Q(x, y) = Cx + Dy + G(x, y)$

where $F(x, y)$ and $G(x, y)$ denote two small functions (necessarily continuous) of order greater than ρ, where $\rho = \sqrt{(x^2+y^2)}$, as $x \to 0$, $y \to 0$, i.e. such that (*uniformly*)

(57) $\qquad \lim_{\rho \to 0} \frac{F(x, y)}{\rho} = 0, \qquad \lim_{\rho \to 0} \frac{G(x, y)}{\rho} = 0$[***]

---

[*] I. BENDIXSON: 'Sur les courbes définies par des équations différentielles,' *Acta Math.*, **24**, 1–88 (1901). See in particular pp. 35–36.

[**] E. R. LONN: 'Über singuläre Punkte gewöhnlicher Differentialgleichungen', *Math. Z.*, **44**, 507–30 (1938). See in particular § 9.

[***] These conditions imply that the functions P and Q vanish at the origin and possess partial derivatives of the first order at the origin such that $P_x(0, 0) = A$, $P_y(0, 0) = B$, $Q_x(0, 0) = C$, $Q_y(0, 0) = D$.

(3) *The two polynomials $Ax+By$ and $Cx+Dy$ do not differ by a constant factor*, i.e. that

(58) $$AD - BC \neq 0$$

Only for the one case in which the discriminant $\Delta = (A - D)^2 + 4BC$ of the characteristic equation is zero is it convenient to replace (57) by the *stronger* condition that the two ratios $F(x, y)/\rho^{1+\varepsilon}$ and $G(x, y)/\rho^{1+\varepsilon}$, where ε denotes any fixed positive number, remain bounded, i.e. in the Landau* notation,

(59) $$F(x, y) = O(\rho^{1+\varepsilon}), \qquad G(x, y) = O(\rho^{1+\varepsilon}) \qquad (\rho \to 0)$$

Using one or other of these conditions together with the preceding hypotheses we can determine an important number connected with a *cycle*** γ embedded in the vectorial field

$$\mathbf{v} = P(x, y)\mathbf{i} + Q(x, y)\mathbf{j}$$

where $\mathbf{i}$ and $\mathbf{j}$ are unit vectors in the directions of the x- and y-axes respectively. The *Poincaré index n* of the cycle γ is defined to be the increment in the angle made by $\mathbf{v}$ with a fixed direction, divided by 2π, when the cycle γ is traced out in the positive direction. (It is assumed that γ passes through no singular point of the field, i.e. through no point at which $\mathbf{v} = 0$.)

From its definition the Poincaré index n is an integer (positive, negative, or zero) and obviously it remains unaltered by a continuous deformation of γ; it must therefore be *zero* if the cycle may be continuously shrunk to a point without crossing a singular point of the field—as certainly happens when the cycle has no singular point within it. On the other hand, if γ contains a finite number of singular points $A_1, A_2, \ldots, A_h$ within it, it is evident that *n is the sum of the Poincaré indices of these singular points*, where by the *index of the singular point* A_h is meant the index of a cycle containing within it only the one singular point A_h.

This shows that the crux of the matter is the determination of the *index of a singular point*. We shall suppose, for convenience, that the origin is the singular point.

Under the hypotheses employed, the equations (56) establish a *topological* correspondence (i.e. bi-uniform and bi-continuous) between a sufficiently

---

* The symbol $O(\ldots)$ which will be much used in the remainder of the book denotes a quantity whose ratio to that quantity contained within the brackets remains finite in the passage to the limit (here as $\rho \to 0$). We shall less frequently use the symbol $o(\ldots)$ to denote a quantity whose ratio to that contained within brackets tends to zero. Thus the statement $u = o(v)$ is equivalent to $\lim u/v = 0$.

** A *cycle* is a closed curve *topologically* equivalent to a circle, i.e. a closed Jordan curve. In other words a *cycle* is a closed curve which may be continuously deformed into a circle.

small neighbourhood of the origin in the (x, y)-plane and the corresponding neighbourhood of the origin in the (P, Q)-plane, where $P = P(x, y)$, $Q = Q(x, y)$, such that to a sufficiently small circle $P^2 + Q^2 = \varepsilon^2$ of the second plane there corresponds a cycle in the (x, y)-plane containing within it the unique singular point O. But in the second plane, the vector **v** is always directed along the outward-drawn normal to the circle, and therefore the angle made by **v** with a fixed direction increases by 2π when the circle is traced out in the positive sense: the index n of O is therefore $+1$ or -1 according as the topological correspondence preserves the direction of rotation, or not, i.e. according as its *Jacobian* is positive or negative.*

In other words, the Poincaré index of the origin is given by the formula

$$(60) \qquad n = \operatorname{sgn}\left\{\frac{\partial(P, Q)}{\partial(x, y)}\right\}_{x=y=0} = \operatorname{sgn}(AD - BC)$$

This shows (cf. § 9) that at least for the abridged equation** *the index of a col is* -1, *while that of a node, a focus, or a centre is* $+1$; for in these last two cases

$$(A-D)^2 + 4BC = (A+D)^2 - 4(AD-BC) < 0$$

thus implying

$$AD - BC > 0$$

Also from the preceding considerations it follows that *within a closed characteristic of* (46) *there lies at least one singular point which is not a col;* for the Poincaré index of a closed characteristic is necessarily $+1$ since the vector **v** is always tangential to the characteristic itself. Consequently within the closed characteristic must lie at least one singular point, and if there is exactly one singular point it cannot be a col.

12. The node

We now proceed to discuss the various possible singular points of the non-linear equation taking as basis the classification in § 9, and we begin with the case of the node, i.e. the case in which

$$\Delta = (A-D)^2 + 4BC > 0, \qquad AD - BC > 0$$

and therefore the characteristic equation (30) has two real and distinct

---

* See, for example, F. TRICOMI (79), Part II, Chapter V, § 15.
** This restriction may be removed as will be seen in the following two sections.

roots λ_1 and λ_2, both of the same sign. We assume, if necessary by changing t into $-t$ (which implies that A, B, C, D, and therefore $\lambda_1 + \lambda_2 = \frac{1}{2}(A + D)$, all change sign) and interchanging the two indices, that λ_1, λ_2, are *positive* and that

(61) $$0 < \lambda_1 < \lambda_2$$

In the case under consideration the same change of variable as used in similar circumstances in § 9 transforms the differential system into the form

(62) $$\frac{d\xi}{dt} = \lambda_1 \xi + F_1(\xi, \eta), \qquad \frac{d\eta}{dt} = \lambda_2 \eta + G_1(\xi, \eta)$$

where $F_1(\xi, \eta)$ and $G_1(\xi, \eta)$ denote two functions satisfying conditions exactly similar to those satisfied previously by the functions F and G.

For the system (62) the two functions Φ and Ψ previously used now take the forms

$$\Phi(\theta) = \lambda_1 + (\lambda_2 - \lambda_1)\sin^2\theta, \qquad \Psi(\theta) = \tfrac{1}{2}(\lambda_2 - \lambda_1)\sin 2\theta$$

By introducing polar coordinates* system (62) may be written as

(63) $$\begin{cases} \dfrac{1}{\rho}\dfrac{d\rho}{dt} = \lambda_1 + (\lambda_2 - \lambda_1)\sin^2\theta + E_1(\rho, \theta) \\[2mm] \dfrac{d\theta}{dt} = \tfrac{1}{2}(\lambda_2 - \lambda_1)\sin 2\theta + E_2(\rho, \theta) \end{cases}$$

where, as in the preceding section, $E_1(\rho, \theta)$ and $E_2(\rho, \theta)$ are two functions tending uniformly to zero as $\rho \to 0$.

These equations show, first, that *every characteristic which approaches sufficiently close to the origin must necessarily approach indefinitely close to it*, or in other words that a characteristic passing through a point P_0 whose radial vector ρ_0 is sufficiently small must be such that on it ρ tends to zero (as $t \to -\infty$).

It is obvious from (61) that

$$\lambda_1 \le \lambda_1 + (\lambda_2 - \lambda_1)\sin^2\theta \le \lambda_2$$

Consequently, if ρ^* is a positive number such that for $\rho < \rho^*$

$$|E_1(\rho, \theta)| < \frac{\lambda_1}{2}$$

---

* Clearly ρ and θ now denote polar coordinates in the $\xi\eta$-plane, not the xy-plane. Further transformations of this kind which may arise will be assumed.

then, by the first equation (63), at any point within the circle of radius ρ^* with centre at the origin we have

$$\frac{\lambda_1}{2} < \frac{1}{\rho}\frac{d\rho}{dt} < \lambda_2 + \frac{\lambda_1}{2} < \frac{3\lambda_2}{2}$$

i.e.

$$\frac{\lambda_1}{2} < \frac{d\log\rho}{dt} < \frac{3\lambda_2}{2}$$

from which, on integrating from the value t_0 of t (to which corresponds a certain $\rho_0 \leq \rho^*$) to a general $t < t_0$, it follows that

$$\frac{\lambda_1}{2}(t-t_0) > \log\rho - \log\rho_0 > \frac{3\lambda_2}{2}(t-t_0)$$

i.e.

(64) $$\rho_0 e^{\frac{\lambda_1(t-t_0)}{2}} > \rho > \rho_0 e^{\frac{3\lambda_2(t-t_0)}{2}}$$

But the exponential on the left evidently tends to zero as $t \to -\infty$; therefore

(65) $$\lim_{t \to -\infty} \rho = 0$$

This is sufficient, as we can apply corollary III of Bendixson's theorem in which $\Psi(\theta)$ is not of constant sign, as here, for example,

$$\Psi\left(\frac{\pi}{4}\right) = \tfrac{1}{2}(\lambda_2 - \lambda_1) > 0, \qquad \Psi\left(\frac{3\pi}{4}\right) = -\tfrac{1}{2}(\lambda_2 - \lambda_1) < 0$$

to conclude that in the case under consideration *the origin is a node whose privileged directions* (i.e. the final directions of the tangents to the characteristics which end there) *are those of the ξ- and η-axes* for which $\Psi(\theta) = 0$.

Rather less easy is the discussion of the case in which, while (58) remains valid, we have

(66) $$\Delta = (A-D)^2 + 4BC = 0$$

Here the two roots λ_1 and λ_2 coincide, and since the function $\Psi(\theta)$ is now of constant sign we may no longer apply corollary III.

In this case, we suppose that $\lambda_1 = \lambda_2 > 0$ and further that in the first instance $A - D \neq 0$ (which, as we have already seen, implies that $B \neq 0$, $C \neq 0$). It is convenient to use the same linear substitution (44) as was used in §9, thus transforming the given system into

(67) $$\frac{d\xi}{dt} = \lambda_1 \xi + F_2(\xi, \eta), \qquad \frac{d\eta}{dt} = \lambda_1(\xi + \eta) + G_2(\xi, \eta)$$

where F_2 and G_2 denote two functions satisfying conditions analogous to those satisfied by the functions F and G. On introducing polar coordinates, we obtain from (67)

(68)
$$\begin{cases} \dfrac{1}{\rho}\dfrac{d\rho}{dt} = \lambda_1(1+\tfrac{1}{2}\sin 2\theta)+E_3(\rho,\theta) \\ \dfrac{d\theta}{dt} = \lambda_1\cos^2\theta+E_4(\rho,\theta) \end{cases}$$

where E_3 and E_4 denote two new functions of ρ and θ which tend uniformly to zero as $\rho\to 0$. In this actual case $\Psi(\theta)=\lambda_1\cos^2\theta$ and is therefore, as has been already pointed out, a function of constant sign.

From the first of equations (68) follow consequences exactly analogous to those already deduced from the first equation in (63); for if ρ^* is a positive number such that within the circle $\rho\leq\rho^*$

$$|E_3(\rho,\theta)|<\tfrac{1}{4}\lambda_1$$

then at all points within this circle

$$-\tfrac{3}{4}\lambda_1<\tfrac{1}{2}\lambda_1\sin 2\theta+E_3(\rho,\theta)<\tfrac{3}{4}\lambda_1$$

and consequently

(69) $$\tfrac{1}{4}\lambda_1<\dfrac{1}{\rho}\dfrac{d\rho}{dt}<\tfrac{7}{4}\lambda_1$$

so that on integrating between any t_0, to which corresponds a value ρ_0 of ρ less than ρ^*, and a general $t<t_0$, we deduce

(70) $$\rho_0 e^{\tfrac{1}{4}\lambda_1(t-t_0)}>\rho>\rho_0 e^{\tfrac{7}{4}\lambda_1(t-t_0)}$$

Therefore, exactly as in the preceding case, we deduce that *the radial vector ρ tends to zero as $t\to -\infty$*.

In this case however greater difficulties arise in showing, by means of the second equation of (68), that as in the case of the abridged equation the origin is a *node* and not a *focus*. The difficulties are not just in method alone, as under the hypotheses used both possibilities may well arise.* If however,

---

* O. PERRON: 'Über die Gestalt der Integralkurven einer Differentialgleichung erster Ordnung in der Umgebung eines singulären Punktes,' *Math. Z.*, 15, 121–46 (1922), and 16, 273–95 (1923). The interesting example

$$\dfrac{dx}{dt}=-x+\dfrac{y}{\log\rho}, \qquad \dfrac{dy}{dt}=-y-\dfrac{x}{\log\rho}$$

is considered, in which $A=-1, B=0, C=0, D=-1$ and therefore $\Delta=0$; nevertheless this system possesses a *focus* at the origin.

See also E. R. LONN, *ibid.*, in which sufficient conditions to distinguish the singularities are given under even more general conditions ($m>1$).

§ 12]   *The node*   59

as has already been pointed out, we replace the conditions (57) by the more restrictive conditions (59) or even add the one further condition* that

(71) $$E_4(\rho, \theta) = o[(\log \rho)^{-2}] \qquad (\rho \to 0)$$

then the ambiguity disappears and the origin is certainly a *node*.

To prove this we begin with the remark that $\sin^2 \theta \leq \theta^2$; also, the second equation (68) is

$$\frac{d\theta}{dt} = \lambda_1 \cos^2 \theta + E_4(\rho, \theta) = \lambda_1 \sin^2\left(\theta \pm \frac{\pi}{2}\right) + E_4(\rho, \theta)$$

Thus if h is a positive number as small as we please, it follows from (71) that for ρ less than a suitable $\bar{\rho} < \rho^*$

(72) $$\frac{d\theta}{dt} < \lambda_1 \left(\theta \pm \frac{\pi}{2}\right)^2 + \frac{h}{\log^2 \rho}$$

From (72) and from the left-hand side of the inequalities (69), we have

$$\frac{d\rho}{dt} > \tfrac{1}{4}\lambda_1 \rho$$

so that, on dividing term by term, we deduce

$$\frac{d\theta}{d\rho} < \frac{1}{\rho}\left[4\left(\theta \pm \frac{\pi}{2}\right)^2 + \frac{4h/\lambda_1}{\log^2 \rho}\right]$$

i.e.

$$\frac{d(\theta \pm \pi/2)}{d\rho} < \frac{1}{\rho}\left[4\left(\theta \pm \frac{\pi}{2}\right)^2 + \frac{a}{\log^2 \rho}\right]$$

where a denotes the positive number $4h/\lambda_1$ which, as h is arbitrary, is also arbitrary. We are thus led to consider the differential equation

$$\frac{d\phi}{d\rho} = \frac{1}{\rho}\left[4\phi^2 + \frac{a}{\log^2 \rho}\right]$$

of which (as may be immediately verified) a particular integral which tends to zero as $\rho \to 0$ is given by

$$\phi = \frac{b}{\log \rho}$$

---

* Except in the special case considered on p. 61.

provided that a and b are connected by the relation

(73) $$4b^2+b+a = 0$$

It is easily seen that if a differentiable function $Y(x)$ is such that in a certain interval

$$\frac{dY}{dx} < f(x, Y)$$

its graph cannot cross more than once, and then *only in a specified direction*, any integral curve of the differential equation

$$\frac{dy}{dx} = f(x, y)$$

since if for $x = x_0$, $y = Y = y_0$, then

$$\left(\frac{d(Y-y)}{dx}\right)_{x=x_0} = \left(\frac{dY}{dx}\right)_{x=x_0} - f(x_0, y_0) < 0$$

and therefore the continuous function $Y - y$ must pass from *positive to negative* values when x increases through the value x_0.*

It follows from this that if b, for example, takes the value $b = -1/8$ (to which corresponds, from (73), $a = 1/16$) the two curves** represented by the equations in polar coordinates

$$\theta = \frac{\pi}{2} - \frac{1}{8\log\rho}, \qquad \theta = -\frac{\pi}{2} - \frac{1}{8\log\rho}$$

and indicated by the letters C_1 and C_2 in figure 10 cannot be crossed more than once for decreasing ρ, and then only in the direction indicated by the arrows,*** by the characteristics of the system we are considering.

On the other hand, since $\Psi(\theta) = \lambda_1 \cos^2\theta$ is always positive except for $\theta = \pm\pi/2 + 2n\pi$, the parts nearest the origin of all lines through the origin (with the exception of the η-axis) cannot be crossed by the characteristics except in the direction of decreasing θ, since from the second equation in (68) it follows that $d\theta/dt$ has the same sign as $\Psi(\theta)$ for sufficiently small

---

* The impossibility of repeated crossings is easily deduced by supposing that x_0 and x_1 are two consecutive zeros of the continuous function $Y - y$ and deriving contradictory conclusions about the sign of $Y - y$ in the interval (x_0, x_1).

** It would appear much simpler to consider the two half-lines $\theta = \theta_0 =$ constant. ($\pi/2 < \theta_0 < \pi$) and $\theta = \theta_0 + \pi$; but in fact the characteristics cross these lines in the same directions as they cross the ξ-axis.

*** Since the difference $\theta \mp \pi/2 + 1/8 \log \rho$ must pass from *negative* values to *positive* values, for decreasing ρ, i.e. the difference must be increasing in the vicinity of a point at which it vanishes.

§ 12]   *The node*   61

values of ρ. In particular, fixing attention on the special values $\theta = 0$ and $\theta = \pi$, we see that the portion of the ξ-axis within the circle $\rho \leq \bar{\rho}$, where $\bar{\rho}$ is a positive number sufficiently small that for $\rho \leq \bar{\rho}$

$$|E_4(\rho, \theta)| < \lambda_1$$

can be crossed by the characteristics for decreasing t only in the direction shown by the arrows in figure 10. But this shows that a characteristic once having entered into either of the two shaded regions of figure 10 cannot leave it; hence there are no characteristics winding round the origin in the form of a spiral. The origin is therefore a *node*, as we wished to show.

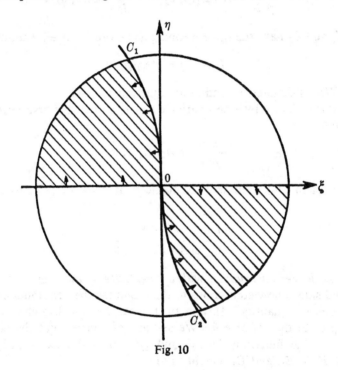

Fig. 10

There remains still to consider the sub-case in which $A - D = 0$, and consequently $BC = 0$, which as we already know may be divided into the further three sub-cases

(a) $B = 0$, $C \neq 0$;   (b) $B \neq 0$, $C = 0$;   (c) $B = C = 0$

However in the subcases (a) and (b)* there is nothing further to add as the sole difference as compared with the case $A - D \neq 0$ is that the given

---

* It is sufficient to consider only the first case (see the note on p. 41).

differential system is now transformed into (67) by the linear substitution $Cx = A\xi$, $y = \eta$ in place of the substitution (44).

In the sub-case (c) we have

$$H(x, y) = Ax, \qquad K(x, y) = Ay$$

and consequently *the polynomial $M(x, y)$ appearing in the theorem of Bendixson is identically zero*, i.e. $\Psi(\theta) \equiv 0$ while $\Phi(\theta) \equiv A \neq 0$.* On introducing polar coordinates, as before, the system becomes

(74) $$\frac{1}{\rho}\frac{d\rho}{dt} = A + E_1(\rho, \theta), \qquad \frac{d\theta}{dt} = E_2(\rho, \theta)$$

where E_1 and E_2 have the same meaning as before; hence by condition (59)

(75) $$E_2(\rho, \theta) = \rho^\varepsilon E_2^*(\rho, \theta)$$

where $E_2^*(\rho, \theta)$ denotes a continuous function as $\rho \to 0$.

We now replace ρ by a new variable $r = \rho^\varepsilon$ so that the first equation (74) takes the form

$$\frac{1}{\varepsilon \rho^\varepsilon}\frac{dr}{dt} = A + E_1(\rho, \theta)$$

and the system (74) itself may be reduced, in view of (75), to the single equation

(76) $$\frac{d\theta}{dr} = \frac{E_2^*(r^{1/\varepsilon}, \theta)}{\varepsilon[A + E_1(r^{1/\varepsilon}, \theta)]}$$

which is no longer an equation with a singularity at the origin. In fact the right-hand side is now continuous at the origin (as the denominator reduces to the non-zero quantity εA), and further satisfies the Lipschitz condition with respect to the variable θ. We assume, of course, that the conditions imposed on the functions $P(x, y)$ and $Q(x, y)$, and consequently on the functions F, G, E_1, and E_2 remain valid.

By the fundamental theorem of Chapter I it follows that equation (76) admits one and only one integral $\theta = \theta(r)$ satisfying the arbitrary initial condition $\theta = \theta_0$ for $r = 0$; therefore *in this special case there is one and only one characteristic reaching the origin in an arbitrarily fixed direction $\theta = \theta_0$*. The origin is therefore a *star-shaped node*.

Lastly it should be noted that if the positive direction of the variable t has an intrinsic significance (as, for example, in a mechanical problem t may denote time) we may then distinguish between *stable nodes* and *unstable*

---

* In this case $AD - BC = A^2$, which implies that $A \neq 0$, by (58).

nodes, according as a variable point on a characteristic which passes through the node approaches the node for increasing t (i.e. for $t \to +\infty$) or for decreasing t (i.e. for $t \to -\infty$).

It is easily seen in the above work that the node under consideration is unstable if $A + D > 0$ and is stable if $A + D < 0$.

13. The focus and the col

We discuss first the case in which

(77) $$(A-D)^2 + 4BC < 0$$

and therefore the two roots λ_1 and λ_2 of the characteristic equation are complex conjugates

$$\lambda_1 = \mu + iv, \quad \lambda_2 = \mu - iv \qquad (v \neq 0)$$

We have already seen that in this case it is convenient to make the linear substitution

$$\begin{cases} X = Cx + (\mu - A)y \\ Y = vy \end{cases}$$

which transforms the system (46) into one of the form

$$\begin{cases} \dfrac{dX}{dt} = \mu X - vY + F_3(X, Y) \\ \dfrac{dY}{dt} = vX + \mu Y + G_3(X, Y) \end{cases}$$

where F_3 and G_3 denote two functions satisfying conditions exactly analogous to those satisfied previously by the functions F and G. Again introducing polar coordinates, i.e. writing

$$X = \rho \cos \theta, \quad Y = \rho \sin \theta$$

and noting that $\Phi(\theta) \equiv \mu$, $\Psi(\theta) \equiv v$, we are led to study the system

(78) $$\frac{1}{\rho}\frac{d\rho}{dt} = \mu + E_5(\rho, \theta), \quad \frac{d\theta}{dt} = v + E_6(\rho, \theta)$$

where E_5 and E_6 denote two new functions tending uniformly to zero as $\rho \to 0$.

Provided that we exclude the case in which $\mu = 0$, i.e. in which $A + D = 0$*, and assume that $\mu > 0$,** we can deduce from the first equation (78) exactly as in the preceding section that *in a sufficiently restricted neighbourhood of the origin ρ always decreases as t decreases and tends to zero as $t \to -\infty$.*

In fact, if ρ^* is a positive number such that within the circle $\rho \leq \rho^*$

$$|E_5(\rho, \theta)| < \tfrac{1}{2}\mu$$

we have

(79) $$\tfrac{1}{2}\mu < \frac{1}{\rho}\frac{d\rho}{dt} < \tfrac{3}{2}\mu$$

an inequality exactly analogous to that in (69) of the preceding section and from which we can deduce corresponding corollaries.

This time however *the characteristics, instead of ending in the origin with definite tangents, wind around the origin in the form of spirals,* so that the origin is a *focus*.

Since for this case the equation $\Psi(\theta) = 0$ has no real roots we can apply the first corollary of the Bendixson theorem (see page 52).

It should be noted that the preceding result is valid under the condition that $A+D$ is not zero. If $A+D = 2\mu = 0$, then (79) is no longer valid and we can no longer assert that ρ tends to zero as $t \to -\infty$. Thus it is not only the method of proof which is no longer applicable but the result itself may fail—as is seen for the abridged equation of the preceding section which in the case $A+D = 0$ does not have a *focus* but has a *centre* at the origin.

There remains therefore the difficulty that these two types of singularities (which when the functions are not analytic are not even the only possibilities) *cannot be distinguished by means of criteria based solely on consideration of the linear terms in P and Q*, i.e. of the four constants A, B, C, D; this is clearly shown by the following example of two systems:

(80) $$\begin{cases} \dfrac{dx}{dt} = y + 2y^3 \\ \dfrac{dy}{dt} = -x - 2x^3 \end{cases}$$
(81) $$\begin{cases} \dfrac{dx}{dt} = y + x(x^2 + y^2) \\ \dfrac{dy}{dt} = -x + y(x^2 + y^2) \end{cases}$$

in which the first-degree terms are identical and which yet possess a *centre* and a *focus* respectively at the origin.

---

* Cf. (31), § 9.   ** If, instead, $\mu < 0$, we would change t into $-t$.

For the first system we have

$$(x+2x^3)dx+(y+2y^3)dy = 0$$

so that the characteristics are the ∞^1 algebraic curves of fourth degree

$$x^2+x^4+y^2+y^4 = c$$

which (as is easily seen) are (for $c > 0$) closed curves enclosing the origin. To integrate the second system it is convenient to introduce polar coordinates $x = \rho \cos \theta$, $y = \rho \sin \theta$, transforming the given equation into

$$\frac{d\rho}{d\theta} = -\rho^3$$

with solution

$$\rho^2 = \frac{1}{2\theta+c}$$

The origin is therefore, as has been already stated, a *focus* of the equation.

It is instructive to compare with these examples the third system

(82)
$$\begin{cases} \dfrac{dx}{dt} = y - x\sqrt{(x^2+y^2)} \sin \dfrac{1}{\sqrt{(x^2+y^2)}} \\ \dfrac{dy}{dt} = -x - y\sqrt{(x^2+y^2)} \sin \dfrac{1}{\sqrt{(x^2+y^2)}} \end{cases}$$

in which the first-degree terms are the same as in the preceding examples but whose other terms are *not analytic* at the origin, i.e. they may *not* be expanded about the origin as power series in x and y. By the introduction of polar coordinates (82) is transformed into

(83)
$$\frac{d\rho}{d\theta} = \rho^2 \sin \frac{1}{\rho}$$

and on replacing $1/\rho$ by z we obtain

$$\frac{dz}{\sin z} = -d\theta$$

whence

$$\log \tan \frac{z}{2} = -(\theta - \theta_0)$$

where θ_0 denotes an arbitrary constant. Thus

$$\tan\frac{z}{2} = \tan\frac{1}{2\rho} = e^{-(\theta-\theta_0)}$$

i.e.

(83') $$\rho = \frac{1}{2\tan^{-1}(e^{-(\theta-\theta_0)})}$$

It follows from (83) that $d\rho/d\theta = 0$ for $\rho = 1/\pi, 1/2\pi, 1/3\pi, \ldots$; hence there are *infinitely many closed characteristics*, viz. the circles whose radii are these quantities and centre the origin.

As regards the other characteristics, we see from (83') that these are spirals which as $\theta \to +\infty$ or $\theta \to -\infty$ indefinitely approach two of these circles, in fact those of radii $1/(2k\pi)$ and $1/[(2k+1)\pi]$ respectively if the inverse tangent is assumed to lie between $k\pi$ and $k\pi + \pi/2$, ($k = 0, 1, 2, \ldots$). For the trivial case $k = 0$ the characteristics move off to infinity as $\theta \to +\infty$.

Figure 11 shows some of the circles of radii $1/\pi, 1/2\pi, 1/3\pi, \ldots$ (which in the terminology of Poincaré are called *limit cycles* of the equation) and some of the characteristics contained between successive pairs of these circles.

Limit cycles will be discussed briefly in § 14. For discussion of further criteria to distinguish the cases of the *centre* and of the *focus* we refer the reader to the work of Perron* already cited and to a paper of Frommer** which deals specifically with this question.

In the case of the *focus* when the positive direction of t has an intrinsic significance we may further distinguish a *stable focus* and an *unstable focus*, the first being obtained as $t \to +\infty$ and the second as $t \to -\infty$. We see, from (79), that there is instability or stability according as $\mu = \frac{1}{2}(\lambda_1 + \lambda_2)$ is positive or negative, i.e. according as $A + D$ is positive or negative, exactly as in the case of the *node*.

Finally we deal with the case in which the abridged equation has a *col*, i.e. when

(84) $$(A-D)^2 + 4BC > 0, \quad AD - BC < 0$$

---

* O. Perron, *ibid*. See in particular **16**, 284 onwards.

** M. Frommer: 'Über das Auftreten von Wirbel und Strudel (geschlossener und spiraliger Integralkurven) in der Umbegung rationaler Unbestimmtheitsstellen,' *Math. Ann.*, **109**, 345–424 (1934).

See also:
N. Bautin: *Dokl. Akad. Nauk. Urss.* (N.S), **24**, 669–72 (1939).
N. A. Sakarnikov: *Prikl. Mat. Mekh.*, **12**, 669–70 (1948); this paper contains corrections and additions to the work of Frommer cited above.

§ 13] *The focus and the col* 67

We wish to show that under the stated conditions the characteristics of system (46) behave in the vicinity of the origin similarly to the characteristics shown in figure 4, i.e. that for the non-linear equation also there is a *col* at the origin.

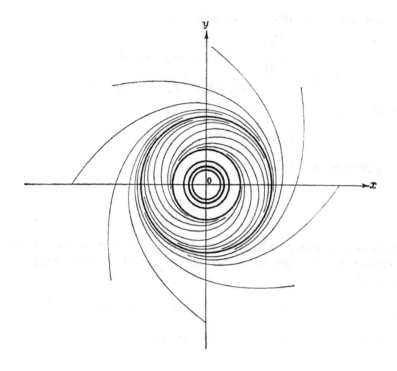

Fig. 11

In this case equations (52) remain valid and may be written in the slightly altered form

(85)
$$\begin{cases} \dfrac{1}{\rho}\dfrac{d\rho}{dt} = \lambda_1 \cos^2\theta + \lambda_2 \sin^2\theta + E_1(\rho, \theta) \\ \dfrac{d\theta}{dt} = \tfrac{1}{2}(\lambda_2 - \lambda_1)\sin 2\theta + E_2(\rho, \theta) \end{cases}$$

But now, while the second of these equations leads to results analogous to those derived in the previous section, the first equation leads to very different conclusions, since as λ_1 and λ_2 are *of different signs* (and we shall suppose $\lambda_1 < 0, \lambda_2 > 0$) the differential coefficient $d\rho/dt$ *changes sign* in the neighbourhoods of the directions $\theta = \pm \alpha$ where $\pm \alpha$ are the values of θ for which

$\lambda_1 \cos^2 \theta + \lambda_2 \sin^2 \theta$ becomes zero, i.e. $\theta = \pm \alpha$ where

(86) $$\tan \alpha = \sqrt{-\frac{\lambda_1}{\lambda_2}}$$

The first equation (85) may be written as

$$\rho \frac{d\rho}{dt} = \lambda_1 \xi^2 + \lambda_2 \eta^2 + (\xi^2 + \eta^2) E_1$$
$$= (\lambda_1 + E_1)\xi^2 + (\lambda_2 + E_1)\eta^2$$

Now let

(87) $$\lambda_1 + E_1 = -\Lambda_1^2, \qquad \lambda_2 + E_1 = \Lambda_2^2$$

(taking account of the signs of λ_1 and λ_2); thus

$$\rho \frac{d\rho}{dt} = (\Lambda_2 \eta - \Lambda_1 \xi)(\Lambda_2 \eta + \Lambda_1 \xi)$$

The locus of points at which $d\rho/dt = 0$ therefore consists of the parts in the vicinity of the origin of the two curves c_1 and c_2 whose equations are respectively

(88) $$\Lambda_2 \eta - \Lambda_1 \xi = 0, \qquad \Lambda_2 \eta + \Lambda_1 \xi = 0$$

The tangents to c_1 and c_2 at the origin are the lines of gradients

$$\pm \lim_{\rho \to 0} \frac{\Lambda_1}{\Lambda_2} = \pm \sqrt{-\frac{\lambda_1}{\lambda_2}}$$

and therefore make with the ξ-axis the angles $+\alpha$ and $-\alpha$ where α is the angle in the first quadrant defined by (86). Hence the curve c_1 lies entirely (at least in the vicinity of the origin) within the first and third quadrants, and the curve c_2 within the second and the fourth quadrants; in both cases the axes themselves are excluded, as the angle θ satisfies on the curve c_1 the conditions

(89) $$\kappa < \theta < \frac{\pi}{2} - \kappa \quad \text{or} \quad \pi + \kappa < \theta < \frac{3\pi}{2} - \kappa$$

and on the curve c_2 the conditions

(89′) $$\frac{\pi}{2} + \kappa < \theta < \pi - \kappa \quad \text{or} \quad \frac{3\pi}{2} + \kappa < \theta < 2\pi - \kappa$$

§ 13] *The focus and the col* 69

where κ is a suitable angle lying between 0 and $\pi/4$. But within the region (89) the function

$$\Psi(\theta) = \tfrac{1}{2}(\lambda_2 - \lambda_1)\sin 2\theta$$

has a *positive* lower bound while within the region (89′) it has a *negative* upper bound; hence, since $E_2(\rho, \theta)$ tends uniformly to zero as $\rho \to 0$, we can determine a positive number $\bar{\rho}$ such that on the parts of the curves c_1 and c_2 within the circle Γ with centre O and radius $\bar{\rho}$ we have respectively

$$\frac{d\theta}{dt} > 0 \text{ (on } c_1\text{)}, \qquad \frac{d\theta}{dt} < 0 \text{ (on } c_2\text{)}$$

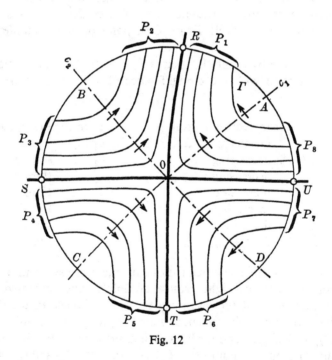

Fig. 12

This implies that the characteristics of the system—*which at all points of c_1 and c_2 have tangents at right angles to the radial vectors*, since at these points $d\rho/dt = 0$—must cross these curves c_1 and c_2 for increasing t in the direction indicated by the arrows in figure 12.

This enables us to trace quite easily the behaviour, either for t increasing or for t decreasing, of the characteristics issuing from points (other than the origin) on the arcs of the two curves c_1 and c_2 within the circle Γ; nor is it difficult to see that, in all cases, at the points just referred to, ρ has a *minimum* value, since

$$\rho \frac{d\rho}{d\theta} = \frac{(\Lambda_2\eta - \Lambda_1\xi)(\Lambda_2\eta + \Lambda_1\xi)}{\frac{1}{2}(\lambda_2 - \lambda_1)\sin 2\theta + E_2}$$

and passes from *negative* values to *positive* values for increasing θ, on crossing the curves c_1 and c_2 specified by the equations (88). Further, as the characteristics issuing from points on c_1 cannot cross c_2 in view of the specified direction at any point of c_2 (for either increasing or decreasing t), while those issuing from points on c_2 cannot cross c_1, we see that on each semi-characteristic issuing from a point on c_1 or on c_2 the radial vector must increase continuously from the initial minimum value, and therefore the semi-characteristic itself must end (if we here confine ourselves to study what happens within the circle Γ) in some point on the circumference of this circle.

If now we denote by points P_1, points P_2, points P_3, ..., points P_8, the points of intersection with the circumference of Γ of the following semi-characteristics,

(1) those issuing for increasing t, from points on c_1 in the first quadrant
(2) ,, ,, ,, increasing ,, ,, ,, c_2 ,, second ,,
(3) ,, ,, ,, decreasing ,, ,, ,, c_2 ,, second ,,
(4) ,, ,, ,, decreasing ,, ,, ,, c_1 ,, third ,,
(5) ,, ,, ,, increasing ,, ,, ,, c_1 ,, third ,,
(6) ,, ,, ,, increasing ,, ,, ,, c_2 ,, fourth ,,
(7) ,, ,, ,, decreasing ,, ,, ,, c_2 ,, fourth ,,
(8) ,, ,, ,, decreasing ,, ,, ,, c_1 ,, first ,,

then since no two distinct characteristics can intersect except in a singular point, it follows that the eight groups of points thus generated will be *separate* in the sense that on describing the circumference in the direction of increasing θ the points P_i are encountered in groups in the order: points P_1, points P_2, points P_3, ..., points P_8, and then again points P_1. It is clear that the four pairs of groups of points (P_2, P_3), (P_4, P_5), (P_6, P_7) and (P_8, P_1) have as separators the four points of intersection A, B, C, D of the curves c_1 and c_2 with the circumference of Γ, these being points in whose neighbourhoods may lie arbitrarily many terminal points P_i of semi-characteristics issuing from points on c_1 and c_2 sufficiently close to the circumference of Γ itself.

With regard to the separators of the other four pairs of groups of points (P_1, P_2), (P_3, P_4), (P_5, P_6), (P_7, P_8), there are two *a priori* possibilities; either they are uniquely determined, i.e. they are four particular points R, S, T, U on the circumference of Γ; or else in at least one of the four cases,

the corresponding groups are separated by a finite arc R'R", S'S", etc., of the circumference of Γ. However it is easily seen that under the conditions we have stated *this second possibility may be excluded*, since it implies that there must be *infinitely many characteristics ending in the origin*, viz. those issuing from points on the arcs R'R", S'S", ... which can have no other end point.*
Perron, to whom we refer the reader for proof,** has shown that under conditions rather more general than those used here there are *exactly four* semi-characteristics which reach the origin. In this particular case these characteristics are those issuing from the four points R, S, T, and U—and the result follows.

It follows therefore from the preceding discussion that when the inequalities (84) are satisfied, the origin is a *col*; this completes the discussion begun at the beginning of § 11. We summarize the results as follows:

Under the hypotheses (1), (2), *and* (3) *enunciated at the beginning of* § 11, *the characteristics of the differential equation*

$$\frac{dy}{dx} = \frac{Q(x, y)}{P(x, y)}$$

in the neighbourhood of the singular point $x = y = 0$, *behave, in general, similarly to the characteristics of the abridged equation*

$$\frac{dy}{dx} = \frac{Cx + Dy}{Ax + By}$$

where

$$P_x(0, 0) = A, \quad P_y(0, 0) = B, \quad Q_x(0, 0) = C, \quad Q_y(0, 0) = D$$

The exceptional cases arise when

$$\Delta = (A-D)^2 + 4BC < 0, \quad A+D = 0$$

for which the abridged equation has a CENTRE, *while the non-linear equation may have a* FOCUS *or even a singularity having in part the features of one and in part the features of the other type of singular point; or when* $\Delta = 0$ *when, to ensure that the non-linear equation* (*like the abridged equation*) *shall have a* NODE, *it is necessary to replace* (57) *by the more restrictive conditions*

$$P(x, y) = Ax + By + O(\rho^{1+\varepsilon}), \quad Q(x, y) = Cx + Dy + O(\rho^{1+\varepsilon})$$

---

* For the semi-characteristic issuing from any point R* of the arc R'R" and approaching the origin must end at O with a specified tangent (in some cases a *vertical* tangent), for it cannot cut the curves c_1 and c_2 in any point other than the origin, since otherwise the point R* would belong either to the group P_1 or to the group P_2, contrary to hypothesis.

** O. PERRON, *ibid.*, **16**, 277–80. See also E. R. LONN, *ibid.*, 514.

where $\rho = \sqrt{(x^2 + y^2)}$ and ε is any small positive number; it should be noted that if B and C are not simultaneously zero in place of $O(\rho^{1+\varepsilon})$ we may write

$$o[\rho(\log \rho)^{-2}]$$

We now add a very brief comment on the case so far excluded, in which

(90) $$AD - BC = 0$$

where A, B, C, D are not all zero (for the case in which $A = B = C = D = 0$ reduces to a differential system of the type considered in the theorem of Bendixson, with $m \geq 2$, which has been studied repeatedly in recent work with results analogous to those obtained in the preceding sections*).

The case (90) has also been thoroughly treated in recent years, particularly by K. A. Keil and A. F. Andreev**; their work has shown that the characteristics in this case may behave in a completely different manner from those of any equation we have so far discussed.

As illustration we consider the simple equation with variables separable

(91) $$\frac{dy}{dx} = \frac{3}{2}\frac{x^2}{y}$$

whose general integral is

$$y^2 = x^3 + c$$

There is therefore a unique characteristic (that corresponding to $c = 0$) which passes through the origin and has a cusp there, while the other characteristics behave as indicated in figure 13.

Lastly we mention briefly a question which has been rather neglected in the theory of first-order differential equations. From the equation

(92) $$f(x, y) = c$$

of the *contour lines* of a surface $z = f(x, y)$, by eliminating the constant we obtain the differential equation of first order

$$f_x(x, y)dx + f_y(x, y)dy = 0$$

---

* In addition to E. R. LONN, *ibid.*, and M. FROMMER, *ibid.*, the following are useful:
M. FROMMER: 'Die Integralkurven einer gewöhnlichen Differentialgleichung erster Ordnung in der Umgebung rationaler Unbestimmtheitsstellen," *Math. Ann.*, **99**, 222-72, (1928).
H. FORSTER: 'Über das Verhalten der Integralkurven einer gewöhnlichen Differentialgleichung erster Ordnung in der Umgebung eines singulären Punktes," *Math. Z.*, **43**, 271-320, (1937).

** K. A. KEIL: *Jahresber. Deutsche Math. Ver.*, **57**, 111-32 (1955); A. F. ANDREEV, *Vestnik Leningrad Univ.* **10**, 43-65 (1955). See also G. SANSONE-R. CONTI (48) and S. LEFSCHETZ (33).

i.e.

(93)
$$\frac{dy}{dx} = -\frac{f_x(x, y)}{f_y(x, y)}$$

Conversely, the equation of the characteristics of a first-order equation may in general be written in the form (92), by solving the equation of the corresponding general integral for the constant. The problems of studying the contour lines on a surface $z = f(x, y)$ and that of studying the characteristics of a first-order differential equation would therefore appear equivalent. However the two problems while being evidently connected are not fully

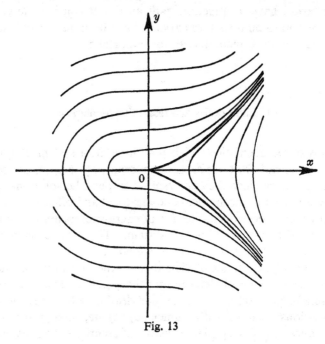

Fig. 13

equivalent, because, if the function $f(x, y)$ satisfies sufficient conditions to justify the inversion of the order of differentiation (in forming derivatives of the second order) the differential equation (93) is less general than those considered in the preceding sections.*

On identifying the preceding functions $P(x, y)$ and $Q(x, y)$ with $f_y(x, y)$ and $-f_x(x, y)$ respectively the inversion theorem for the order of differentiation implies

$$P_x + Q_y = f_{xy} - f_{yx} = 0$$

* For a parallel study of the two problems, with many examples, see G. Bouligand (8). Even here however the question of the equivalence or otherwise of the two problems is not explicitly dealt with.

Therefore at the singular points of (93), i.e. at the points in which f_x and f_y vanish simultaneously, the condition $A + D = 0$ is automatically satisfied—and this is the condition that characterizes the exceptional case in which a *centre* may occur.

This accounts for the fact that such a singularity which is to be regarded as *exceptional* for the characteristics of a first-order differential equation of general type is *usual* for the contour lines on a surface $z = f(x, y)$ of ordinary type; these contour lines possess *centres* at the maximum and minimum points of the function $f(x, y)$.

Naturally assertions of this kind, expressed as *exceptional*, *usual*, and in similar terms, have no precise significance, nor can have any. However we see no reason to omit such remarks as have been made since they help to illustrate mathematical facts that are of importance.

14. Limit cycles and relaxation oscillations

The closed characteristics which may arise in a system of type (9) and which are in general *limit cycles* as defined on page 66, are of considerable physical importance since they correspond to periodic solutions in t; these solutions may occur in mechanical and electrical and other non-conservative systems in which some external source of energy compensates for the energy dissipated by the resistance of the system. The term usually applied to oscillations of this type is *relaxation oscillations*.

From the analytical point of view there is the difficulty that there are no general methods known of determining the possible closed characteristics; all that are available are some conditions dealing with necessity alone, and some conditions dealing with sufficiency alone, and some procedures applicable only to a few special types of differential equations, as we shall see in the following work. Perhaps this is not surprising as the existence of a closed characteristic is not a *local* but a *global* property of the vector field associated with the differential equation, and in all branches of mathematics global properties are, in general, the most difficult to establish.

Since, as we have already seen, the Poincaré index of a cycle is always $+1$ it follows that *a necessary condition for the existence of a closed characteristic in a certain simply-connected domain is that it contains at least one singular point whose Poincaré index equals* $+1$, i.e. *that it contains a* FOCUS *or a* CENTRE *or a* NODE.

In the second place, as the *flux* of the vector

$$\mathbf{v} = P(x, y)\mathbf{i} + Q(x, y)\mathbf{j}$$

across a closed characteristic is necessarily zero as the vector is always tangential to the characteristic itself, we see that *the boundary c of a simply-connected domain D may be a characteristic (necessarily closed) of the differential system* (9) *only if*

$$\text{(94)} \quad \iint_D \operatorname{div} \mathbf{v}\, dx\, dy = \iint_D \left(\frac{\partial P}{\partial x} + \frac{\partial Q}{\partial y}\right) dx\, dy = 0$$

for, supposing that P and Q and their first derivatives are continuous, the flux of $\mathbf{v}$ across the boundary equals the double integral of div $\mathbf{v}$ throughout the domain, with the sign changed.* It follows therefore from (94) that *there can be no closed characteristics entirely contained in a domain in which the quantity $\partial P/\partial x + \partial Q/\partial y$ is of constant sign.*

While the preceding conditions are necessary, a sufficient condition is stated in the following theorem of Bendixson:

If at the frontier of a certain doubly-connected domain D with no singular points within it the vector $\mathbf{v}$ is always directed towards the interior (or towards the exterior) of D, then there exists within D at least one closed characteristic.

The theorem is a consequence of the fact that if the vector $\mathbf{v}$ is always directed inwards (or outwards) to D the characteristic passing through any point P_0 of D when continued in the direction of increasing t (decreasing t) can never leave D, and therefore if it is not a closed curve it must necessarily be of spiral form.

With this in mind we introduce for greater clarity a system of polar coordinates (ρ, θ) with origin O within the lacuna in the domain D and with the polar axis meeting the internal and external boundaries of D in the points A and B respectively. Consider now the continuation of the characteristic c issuing from a certain point $P_0(\rho_0, 0)$ on the segment AB (figure 14); we denote by ρ_1 the value assumed by ρ at the point P_1 (on the polar axis) of the characteristic c when θ has increased (or diminished) by 2π. The three following possibilities may clearly occur:

(1) $\rho_1 = \rho_0$.

(2) There are certain values of ρ_0 for which $\rho_1 > \rho_0$ (as shown in the figure) and other values for which $\rho_1 < \rho_0$.

(3) For all values of ρ_0 in the interval (a, b) determined by the radial vectors of A and B, $\rho_1 > \rho_0$ or alternatively $\rho_1 < \rho_0$.

In the first case the characteristic c is itself a cycle and no more need be said.

In the second case the existence of at least one cycle is immediately

---

* See, for example, F. TRICOMI (79), Part II, Chapter VI, § 6.

established since the difference $\rho_1 - \rho_0$ which is evidently a continuous function of ρ_0 assumes both positive and negative values in the interval $a \leqq \rho_0 \leqq b$.

In the third case, assuming (for example) that $\rho_1 > \rho_0$ and that $\rho_2, \rho_3, \rho_4, \ldots$, are the successive values of ρ when c cuts the polar axis for the second, third, ..., times, we note that the sequence $\rho_1, \rho_2, \rho_3, \ldots$, must tend to some limit $\rho^* \leqq b$, since it is both increasing and bounded (all the ρ_h are not greater than b). Now putting $\rho_0 = \rho^*$ it follows that $\rho_1 = \rho_0 = \rho^*$ since the difference $\rho_1 - \rho_0$ tends to zero when ρ_0 tends to ρ^* through the preceding sequence of values.

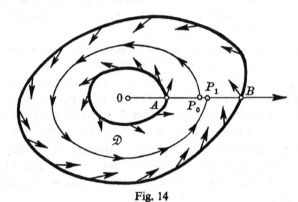

Fig. 14

The preceding criteria have the merit of simplicity, but they seldom suffice alone to decide the existence or otherwise of cycles.* For example, consider the important case of the *generalized equation of Liénard:*

$$(95) \qquad \frac{d^2x}{dt^2} + f(x)\frac{dx}{dt} + g(x) = 0$$

where $f(x)$ and $g(x)$ are two given functions satisfying conditions which will be stated shortly. We note that this equation reduces to the actual equation of Liénard when $g(x) = x$, and if, in addition, $f(x) = \mu(x^2 - 1)$ where μ is a positive constant, the equation is the well-known *equation of van der Pol:*

$$(96) \qquad \frac{d^2x}{dt^2} + \mu(x^2 - 1)\frac{dx}{dt} + x = 0$$

* Recent research by I. G. PETROVSKI and E. M. LANDIS has discussed the maximum number of cycles (exceptional cases omitted) that may occur for a system of the type (46) when P, Q are both polynomials of degree n, say. Even for the case $n = 2$ many difficulties arise in establishing that the maximum number is *three*. See I. G. PETROVSKI and E. M. LANDIS, *Dokl. Akad. Nauk. SSSR* (N.S.), **102**, 29–32 (1955); *Mat. Sbornik* **37** (79) 209–50 (1955).

This equation was the first non-linear equation of real physical importance in which the existence of a periodic solution was established originally by experimental means.*

We now propose, following a topological method due to N. Levinson and O. K. Smith,** to prove that a first-order equation of the form (95) has *one and only one closed characteristic, under the three following conditions:*

(1) The continuous functions $f(x)$ and $g(x)$ are defined on the entire x-axis, $f(x)$ being an *even* function and $g(x)$ an *odd* function such that $xg(x) > 0$ for all $x \neq 0$.

(2) If

(97) $$F(x) = \int_0^x f(\xi)\, d\xi$$

then the *odd* function $F(x)$ has exactly one positive zero at $x = \alpha$ and is always negative for $0 < x < \alpha$.

(3) The function $F(x)$ is positive, non-decreasing for $x > \alpha$, and tends to $+\infty$ as x tends to $+\infty$.

Since, employing (97),

$$\frac{d^2x}{dt^2} + f(x)\frac{dx}{dt} = \frac{d}{dt}\left[\frac{dx}{dt} + F(x)\right]$$

on putting

(98) $$\frac{dx}{dt} + F(x) = y$$

equation (95) is transformed into the system of type (9),

(99) $$\frac{dx}{dt} = y - F(x), \qquad \frac{dy}{dt} = -g(x)$$

or into the single equation

(100) $$\frac{dy}{dx} = -\frac{g(x)}{y - F(x)}$$

---

* *Experimentally* in a double sense ... as van der Pol recognized the existence of a closed cycle in the phase space associated with (96) partly from his experiences with certain electrical circuits governed by the equation and partly by plotting the field of tangential directions by the method of isoclines.

** See N. Minorsky (72), pp. 108 onwards.

which under the above hypotheses, has the unique singular point $x = y = 0$, for which (in the usual notation)

$$AD - BC = g'(0) > 0$$

The singular point cannot therefore be a col, and must have Poincaré index equal to $+1$.

As both functions $g(x)$ and $F(x)$ are odd, it is easily seen that *for this equation the field of directions in the (x, y)-plane is symmetrical with respect to the origin*, as the right-hand side of equation (100) remains unaltered when x and y are simultaneously changed into $-x$ and $-y$.

Second, we see that the y-axis (which is the locus of points at which $dy/dx = 0$) and the curve F with equation $y = F(x)$ (which is the locus of points at which $dy/dx = \infty$) divide the plane into four regions, denoted by the numbers I, II, III, IV in figure 15, in which the coordinates (x, y) of points on any characteristic of the equation behave as follows, for increasing t:

(I) x increasing, y decreasing;   (II) x increasing, y increasing;
(III) x decreasing, y increasing;   (IV) x decreasing, y decreasing.

It is clear that the curve F, the y-axis, and the line r drawn parallel to it through any point C whatsoever on the part of the curve F to the right of the origin cannot be crossed (for increasing t) by characteristics of the equation *except* in the direction indicated by the arrows in figure 15, in view of the signs of the right-hand sides of equations (99).

Of the two semi-characteristics issuing from the point C mentioned above (to which we suppose corresponds the value t_0 of t), that corresponding to values of t greater than t_0 will cross into the part of region IV to the left of r and will remain there until it cuts the y-axis in some point B lying below O. On the other hand the other semi-characteristic will cross, for decreasing t, from $t = t_0$, into the part of region I to the left of r and will remain there until it cuts the y-axis in some point A lying above O.

For our purposes it is important to compare the lengths of the two segments OA and OB, as it is easily seen that *the necessary and sufficient condition that we obtain a closed curve on continuing the two preceding semi-characteristics beyond the points A and B into the half-plane $x < 0$ is that OA = OB*.

The condition is *sufficient*, since if OA = OB the semi-characteristic CA may be continued beyond A by simply reflecting the semi-characteristic BC in the origin, and the semi-characteristic CB by reflecting the semi-characteristic AC in the origin; this necessarily gives rise to a closed curve.

Conversely, the condition is also *necessary* since any closed characteristic (which must necessarily contain the origin O in its interior) must be symmetrical with respect to the origin O, and therefore must cut the y-axis (or the x-axis, or any straight line passing through the origin) in two points A and B such that OA = OB.*

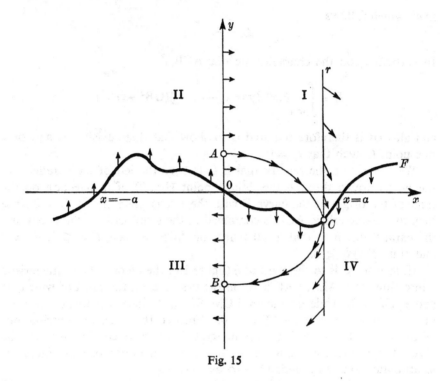

Fig. 15

It therefore follows that the existence and uniqueness of a closed characteristic of (100) will be established if we can show that there is exactly one point C on the arm of the curve F to the right of the y-axis such that on the characteristic $c \equiv \mathrm{ACB}$ we have OA = OB.

We introduce the function

$$\lambda(x, y) = \tfrac{1}{2}y^2 + \int_0^x g(\xi)\, d\xi$$

---

* Alternatively, by employing the symmetry with respect to the origin O of the given closed characteristic c_1 we obtain a second closed characteristic c_2; as these characteristics cannot cut except in a singular point either c_2 must lie entirely inside c_1 or c_1 entirely inside c_2. Hence, if R_1 and R_2 denote the greatest distances measured along any line r which passes through the origin from O to points of c_1 and of c_2 respectively, $R_2 < R_1$ if c_2 lies entirely within c_1, or $R_1 < R_2$ if c_1 lies within c_2—and both of these relations are impossible since the symmetry property implies $R_1 = R_2$.

and note that on any characteristic

$$\frac{d\lambda}{dt} = y\frac{dy}{dt} + g(x)\frac{dx}{dt} = -F(x)g(x) = F(x)\frac{dy}{dt}$$

from which follows

$$d\lambda = F(x)\,dy$$

In particular, on the characteristic $c \equiv ACB$,

$$I_c \equiv \int_{ACB} F(x)\,dy = \lambda_B - \lambda_A = \tfrac{1}{2}(OB^2 - OA^2)$$

and all that is therefore required is to show that there exists one and only one point C such that $I_c = 0$.

We begin with the remark that if C falls to the left of the parallel h to the y-axis drawn through the unique point $H(\alpha, 0)$ of intersection on the right of the origin of the curve F with the x-axis, then clearly $I_c > 0$ since then the characteristic ACB is contained in the strip between the y-axis and the parallel line h so that at all points on ACB we have $F(x) \leq 0$, $dy < 0$ and thus $F(x)dy \geq 0$.

If, instead, C is to the right of h (figure 16), then h cuts the characteristic c into three arcs AL, LCM, MB, on the first and on the third of which, as above, $Fdy \geq 0$, while on the arc LCM $Fdy \leq 0$, since at points to the right of h we have $dy < 0$ but $F(x) > 0$. Further the positive contributions of the arcs AL and MB to the integral I_c decrease continuously as the abscissa x_0 of C increases, while the negative contribution of LCM increases in absolute value and tends to $-\infty$ as $x_0 \to +\infty$.

To prove this we consider in addition to the characteristic $c \equiv ACB$ a second characteristic $c' = A'C'B'$ obtained by giving a positive increment (small or large) to the abscissa x_0. This second characteristic can never intersect the former one; also, in view of (100),

$$d\lambda = -\frac{F(x)g(x)}{y - F(x)}\,dx.$$

Thus, in an obvious notation,

$$\int_{A'}^{L'} F\,dy - \int_{A}^{L} F\,dy = -\int_{0}^{a}\left[\frac{1}{y_{c'} - F(x)} - \frac{1}{y_c - F(x)}\right]F(x)g(x)\,dx < 0$$

and similarly

$$\int_{M'}^{B'} F\,dy - \int_{M}^{B} F\,dy < 0$$

§ 14] *Limit cycles and relaxation oscillations* 81

As for the (negative) contribution of the arc LCM it is easily seen that this increases in absolute value as we pass from c to c' since, if L_1 and M_1 are the points of c' having the same ordinates as L and M,

$$-\int_{L'}^{M'} F\,dy + \int_{L}^{M} F\,dy > -\int_{L_1}^{M_1} F\,dy + \int_{L}^{M} F\,dy \geqq 0$$

for at two points with the same ordinate y the value of the F appearing in the last integral is not greater than that of the F appearing in the second last integral. (This follows since $dy < 0$.)

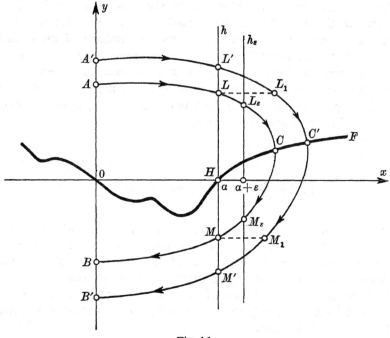

Fig. 16

Finally we note that the contribution of the arc LCM tends to $-\infty$ as $x_0 \to +\infty$; for if ε be a positive number as small as we please, and L_ε and M_ε the intersections of c with the line h_ε with equation $x = \alpha + \varepsilon$ (i.e. with the line h moved a distance ε to the right) the theorem of mean value yields

$$-\int_{L}^{M} F(x)\,dy > -\int_{L_\varepsilon}^{M_\varepsilon} F(x)\,dy = [F\alpha + \varepsilon + \theta(x_0 - \alpha - \varepsilon)](y_{L_\varepsilon} - y_{M_\varepsilon})$$

where θ indicates a suitable number lying between 0 and 1. But $y_{L_\varepsilon} - y_{M_\varepsilon}$ does not tend to zero as $x_0 \to +\infty$ since this difference is non-decreasing (as two characteristics cannot intersect), while $F[\alpha + \varepsilon + \theta(x_0 - \alpha - \varepsilon)]$ tends to $+\infty$ as $x_0 \to \infty$*, by assumption (3); the result then follows.

We have therefore shown that, as the abscissa x_0 of C increases, the integral I_c which is initially positive begins to decrease from a certain point onwards (attained for $x_0 > \alpha$), becomes negative, and tends to $-\infty$ as $x_0 \to +\infty$. There exists therefore one and only one value of x_0, i.e. one and only one point C for which $I_c = 0$, i.e. OA = OB, and we conclude that *under the conditions enunciated above the generalized Liénard equation* (95) *possesses one and only one periodic solution.*

Among recent work on the Liénard equation and its generalizations has been that of D. Graffi** who determined a lower bound for the interval between two successive zeros of any integral whatsoever of the equation; also that of G. Sansone*** who established some criteria for existence and uniqueness of closed characteristics under conditions more general than those used here, and derived a comparison theorem for the actual Liénard equation on which may be based a method of approximation to the period of the relaxation oscillation.

15. Periodic solutions in the phase space

In several questions, for example some concerned with synchronous electrical motors, we are interested not only in the periodic solutions of non-linear equations of the type considered in the preceding section—which correspond to closed characteristics in the phase space—but also in solutions representable as a sum of periodic terms plus *secular terms* (i.e. terms which increase linearly with the time t), which correspond to periodic solutions in the phase space.

We note, first, that in order that there may exist a periodic solution $y = \phi(x)$ of the first-order differential equation

$$y' = f(x, y)$$

of period ω, or even an infinite number of such solutions, it is not necessary

---

* To show clearly that the argument of F actually does tend to $+\infty$ as $x_0 \to +\infty$ (although θ may be very small) it is sufficient to note that we can take $\varepsilon = (x_0 - \alpha)/2$ which implies $\alpha + \varepsilon = (x_0 + \alpha)/2$.
** D. Graffi: *Mem. Accad. Sc. Bologna*, (9) **7**, 121–29 (1940); also (9) **9**, 83–91 (1942)
*** See G. Sansone: *Ann. Mat.*, (4) **28**, 153–81 (1949); *Rend. Seminario Mat. Torino*, **10**, 155–71 (1950–51).

that the function $f(x, y)$ considered as a function of x be itself periodic of period ω, since the existence of the solution $y = \phi(x)$ affects only the values of $f(x, y)$ at points near to those of the corresponding characteristic. However, if we demand that *all* the solutions of equation (1) be periodic of period ω,* then the function $f(x, y)$ considered as a function of x must itself be periodic of period ω; for if the characteristic issuing from any point (x_0, y_0) is represented by the equation $y = \phi(x)$ where ϕ is a periodic function with period ω, then

$$\phi(x_0+n\omega) = \phi(x_0) = y_0, \qquad \phi'(x_0+n\omega) = \phi'(x_0)$$

which implies

$$f(x_0+n\omega, y_0) = f(x_0, y_0)$$

As x_0 and y_0 are arbitrarily chosen this shows that f is a periodic function of x with period ω.

Conversely, if f is a periodic function of x of period ω, we might think that there must be at least one solution of period ω; this is not true in general, as is shown by the simple example of a linear homogeneous equation

$$y' = A(x)\,y$$

with general integral

(101) $$y = Ce^{B(x)}$$

where C denotes an arbitrary constant and

$$B(x) = \int_{x_0}^{x} A(\xi)\,d\xi$$

The function (101) is periodic (with period ω) if and only if the function $B(x)$ is, and for this it is necessary (as $A(x) = B'(x)$) but not sufficient that $A(x)$ be periodic with period ω; for the periodicity of $B(x)$ we require the additional condition

(102) $$\int_{x}^{x+\omega} A(\xi)\,d\xi = 0$$

Disregarding the case, which may be considered exceptional, in which an equation of the type (1) possesses solutions of period ω while the function f is not periodic in x with period ω, it is evident from what has been said

---

* For simplicity we do not discuss whether the condition '*all* the solutions' can be weakened (as in fact it can). We also assume that the function $f(x, y)$ satisfies all the conditions required for the validity of the fundamental theorem of Chapter I.

above that it is of some importance to derive criteria which answer the following question:

Given an equation of the type (1) *in which* $f(x, y)$ *is a periodic function of x, does it possess solutions which are also periodic and of the same period?*

The conditions imposed on the function $f(x, y)$, in addition to those under which the fundamental theorem of Chapter I is valid, allow us to make this assertion:

The necessary and sufficient condition that to the characteristic $y = \phi(x)$ *of the equation* (1) *issuing from some point* (x_0, y_0) *there corresponds a periodic solution with period* ω, *is*

$$(103) \qquad \phi(x_0 + \omega) = \phi(x_0)$$

The condition (103) which is evidently necessary is also sufficient, since by changing x into $x + \omega$ (which in view of the periodicity of the function f does not alter the given differential equation), we have two functions $y = \phi(x)$ and $y = \phi(x + \omega)$ satisfying both equation (1) and the same initial conditions $y(x_0) = y_0$, where y_0 denotes the common value of both sides of equation (103), and these two solutions must coincide.

From the preceding remarks may be deduced a method, theoretically sound but not always useful in practice in the form in which it is first stated, which answers the question raised above:

Let x_0 *be any value whatsoever of x and consider the integral* $y = \phi(x, \eta)$ *of the given equation which assumes the value* $y = \eta$ *for* $x = x_0$; *to every root* (*if any exist*) *of the equation in* η,

$$(104) \qquad \Phi(\eta) \equiv \phi(x_0 + \omega, \eta) - \eta = 0$$

there corresponds a periodic solution, with period ω, *of the given differential equation, and conversely.*

The usefulness of this criterion is in general limited by the difficulty of finding explicitly the function $\phi(x, \eta)$.

However, as the function $\Phi(\eta)$ is necessarily a continuous function of η (as follows from the theorem on the dependence of integrals on given initial conditions; see p. 12), the preceding method allows us to establish the existence of periodic solutions more easily by showing that there are values of η for which $\Phi(\eta) > 0$ and values of η for which $\Phi(\eta) < 0$—for then there must be at least one value of η for which (104) is satisfied.*

A criterion of existence and uniqueness of considerable practical importance is stated in the following theorem, analogous to the theorem of Bendixson discussed in the preceding section:

---

* F. TRICOMI: 'Periodische Lösungen einer Differentialgleichung erster Ordnung,' *Verh. Int. Math. Kongress Zürich*, **2**, 72–73 (1932).

If the differential equation $y' = f(x, y)$ where f is a periodic function of x with period ω satisfies the conditions demanded for the validity of the fundamental theorem of existence and uniqueness (i.e. there are no singular points) in the whole strip D defined by the inequalities

$$-\infty < x < \infty, \qquad \alpha(x) \leq y \leq \beta(x)$$

where $\alpha(x)$ and $\beta(x)$ are any two periodic functions of period ω (which may possibly be two constants), and if in addition on the curves $y = \alpha(x)$ and $y = \beta(x)$ which make up the boundary of D the vector $\mathbf{v} = \mathbf{i} + f(x, y)\,\mathbf{j}$ is always directed towards the interior of D or always towards the exterior of D, then within the strip D there exists at least one periodic solution $y = y^*(x)$ of the equation with period ω.*

Further, if the function f, considered as a function of y, is always increasing or always decreasing in D, the preceding periodic solution is unique.**

As in the case discussed in the preceding section the theorem hinges on the fact that if the vector $\mathbf{v}$ at any point on the boundary of D is always directed inwards (outwards) relative to the strip D, the semi-characteristic which issues from any point (x_0, y_0) of D for increasing (decreasing) values of x can never leave the strip D.

Consequently, assuming that the vector $\mathbf{v}$ is (as in figure 17) always directed towards the interior of D and in addition that $x_0 = 0$, the semi-characteristic $x \geq x_0$ will meet the successive parallel lines $x = \omega$, $x = 2\omega$, ..., in certain points of D with ordinates indicated respectively by y_1, y_2, ... Also, as both the equation and the strip D are unaltered by any translation parallel to the y-axis of length a multiple of ω, it is clear that the same relation must hold between y_n and y_{n+1} ($n = 1, 2, 3, \ldots$) as between y_0 and y_1, so that if y_n coincides with y_0 then y_{n+1} must coincide with y_1.

---

* In the paper of L. AMERIO: 'Studio asintotico del moto di un punto su una linea chiusa, per azione di forze indipendenti dal tempo', *Ann. Sc. Norm. Sup. Pisa*, (3) 3, 19–57 (1949), Amerio has generalized earlier work of the author of this book; his results to which reference will later be made (p. 87) contain two lemmas (I and II) which together make up a theorem similar to that proved above. Instead of imposing the condition on the vector $\mathbf{v}$ at points on the frontier of D he postulates that there exists at least one semi-characteristic which does not issue from the prescribed strip.

** Evidently it will be sufficient to verify the condition on the vector $\mathbf{v}$ in an interval of width ω, for example in the interval $0 \leq x \leq \omega$. This greatly weakens the restriction that $\alpha(x)$ and $\beta(x)$ be *periodic* functions, reducing it to simply $\alpha(0) = \alpha(\omega)$, $\beta(0) = \beta(\omega)$. Otherwise these conditions may be weakened to the demand that $\mathbf{v}$ satisfies the prescribed conditions also on the segments of the line $x = \omega$ between $\alpha(0)$ and $\alpha(\omega)$ and between $\beta(0)$ and $\beta(\omega)$.

It should be noted that if the strip is bounded by two lines parallel to the x-axis the condition on the vector $\mathbf{v}$ is equivalent to demanding that the function $f(x, y)$ is of constant sign along the two lines, in fact is positive on one line and negative on the other.

We now distinguish the three following cases:

(1) We have $y_1 = y_0$.

(2) There are certain values of y_0 lying between $\alpha(0)$ and $\beta(0)$ for which $y_1 > y_0$ (as in figure 17) and others for which $y_1 < y_0$.

(3) For any value of y_0 in the interval $\alpha(0) < y_0 < \beta(0)$, we always have $y_1 > y_0$ or always $y_1 < y_0$.

In the first and the second cases the existence of at least one periodic solution is evident, in view of the remarks above. In the third case it follows, as in the theorem of Bendixson, that the sequence $y_0, y_1, y_2, \ldots$, tends to

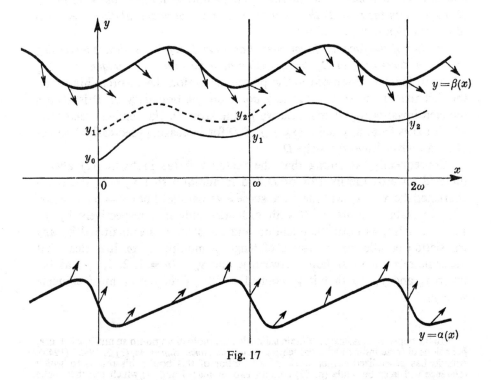

Fig. 17

some limit y^* lying between $\alpha(0)$ and $\beta(0)$, since it is both monotonic and bounded. On writing $y_0 = y^*$, it is evident that to this limit as initial value there corresponds a characteristic for which $y_1 = y_0$ (i.e. a periodic solution) since the difference $y_1 - y_0$ tends to zero as y_0 tends to y^* through the sequence $y_0, y_1, y_2, \ldots$.

Finally, by imposing the additional condition that the function f considered as a function of y is always increasing or always decreasing, we can show the periodic solution derived above and denoted here by $y^*(x)$ to be *unique*; for by integrating the given equation in which $y^*(x)$ is written in place of $y(x)$, from 0 to ω we have

$$\int_0^\omega f[x, y^*(x)]\,dx = y^*(\omega) - y^*(0) = 0$$

so that if there exist two distinct periodic solutions $y_1^*(x)$ and $y_2^*(x)$ then

$$\int_0^\omega f[x, y_1^*(x)]\,dx = \int_0^\omega f[x, y_2^*(x)]\,dx$$

But, on the other hand, either always $y_1^*(x) > y_2^*(x)$ or $y_1^*(x) < y_2^*(x)$, since two distinct characteristics can never intersect in the strip D, as it is without singular points, and hence the first of the two integrals must be greater than the second or be less than the second, according as the first or second of the above inequalities holds, i.e. according as the function $f(x, y)$ increases or decreases with increasing y. Hence the supposition that there may be two distinct periodic solutions $y_1^*(x)$ and $y_2^*(x)$ leads to an absurdity, and therefore there can be no more than one.

This last theorem is useful in studying the differential equation

(105)
$$\frac{dy}{dx} = -\frac{\alpha y + \beta + \sin x}{2y}$$

where α and β are two positive constants. This equation arises in electrical and other important physical problems, and it has been thoroughly discussed by several writers, including the author about thirty years ago,[*] and more recently by L. Amerio.[**]

Here we omit discussion of the most difficult case in which $0 < \beta \leq 1$, and we suppose $\beta > 1$. Since $\beta + \sin x > 0$ for all x, the function

$$f(x, y) = -\frac{\alpha y + \beta + \sin x}{2y}$$

never assumes the indefinite form 0/0; further, this function is *periodic in x* (with period 2π) and is an *increasing function of y* which assumes *values*

---

[*] F. Tricomi: 'Sur une équation différentielle de l'electrotéchnique,' *C. R. Acad. Sci. Paris*, **193**, 635–36 (1931); also 'Integrazione di un'equazione differenziale presentatasi in elettrotecnica,' *Ann. Sc. Norm. Sup. Pisa*, (2), **2**, 1–20 (1933). The principal results of these papers are reported in Andronov and Chaikin (55) pp. 293–300.
  The equation is fully discussed also in Bieberbach (6) and in Sansone-Conti (48).

[**] L. Amerio: 'Determinazione delle condizioni di stabilità per gli integrali di una equazione interessante l'elettrotecnica," *Ann. Mat.*, (4) **30**, 75–90 (1949).

of contrary sign (being negative at the lower extreme and positive at the upper extreme) on the boundary of the strip

(106) $$-\frac{\beta+1}{\alpha} - \varepsilon \leq y \leq -\frac{\beta-1}{\alpha} + \varepsilon$$

where ε is an arbitrarily small positive number. But this implies that at the boundary of the strip (106) the vector $\mathbf{v} = \mathbf{i} + f(x, y)\,\mathbf{j}$ is always directed *outwards* relative to the strip; hence the strip defined by (106) in which we can now even allow $\varepsilon = 0$ contains one and only one periodic solution of the equation (105).

III. Boundary problems for linear equations of the second order

16. Preliminary considerations

In this chapter we shall discuss only *linear differential equations of the second order*, as this can be done with greater simplicity, although many properties which we shall derive for these equations may be easily extended to cover linear equations of any order, and some of the properties may even be extended to non-linear equations. However in this field, as in all others in mathematics, it is helpful to discuss new problems and their inherent difficulties first in the simplest form possible, and later investigate to what extent the results may be generalized.

Given a linear and homogenous* equation of the second order

$$(1) \qquad A(x)y'' + B(x)y' + C(x)y = 0$$

the fundamental theorem of Chapter I establishes that this equation possesses one and only one integral satisfying the arbitrary *initial conditions*

$$(2) \qquad y(x_0) = \alpha, \qquad y'(x_0) = \beta$$

provided that the coefficients $A(x)$, $B(x)$, and $C(x)$ are continuous in the vicinity of the point $x = x_0$, and in addition that $A(x_0) \neq 0$.

Supplementary** conditions of the form (2) are frequently met in practical problems, especially those of mechanics, in which the given data are frequently the initial position and velocity. There are also problems of importance involving supplementary conditions of different type—for example, in connection with the problem of the *brachistochrone*, i.e. the curve traced out by a weight moving in the shortest possible time from one point $A = (a, A)$

---

* In this and subsequent sections we shall consider only *homogeneous* equations as the integration of a linear non-homogeneous equation can be reduced to that of the corresponding homogeneous equation by Lagrange's method of 'variation of parameters'.

** We use the phrase 'supplementary conditions' to describe all additional conditions which may be associated with a differential equation, whether of the form (2) or of another kind. (The German word is *Nebenbedingungen*.)

to a point $B = (b, B)$ in the vertical (x, y) plane,* we require to integrate the (non-linear) equation of second order

$$yy'' + \tfrac{1}{2}(1 + y'^2) = 0$$

together with the *boundary conditions*

(3) $\qquad\qquad\qquad y(a) = A, \qquad y(b) = B$

More generally, boundary conditions take the form

(4) $\qquad hy(a) + h'y'(a) = A, \qquad ky(b) + k'y'(b) = B$

where h, h', k, k', A, B are six given constants such that h and h', and k and k' are not simultaneously zero.**

Problems of this kind are easily dealt with for the *linear* equation, which is the only one we consider at present, provided we know two linearly independent integrals $y_1(x)$ and $y_2(x)$ of the given equation (1).

Every other integral of the equation is of the form $c_1 y_1(x) + c_2 y_2(x)$ where c_1 and c_2 denote arbitrary constants; the possibility or otherwise therefore of satisfying the supplementary conditions (4) depends on the possibility or otherwise of solving for c_1 and c_2 the linear system

(4') $\qquad\qquad \begin{cases} c_1 F_1(a) + c_2 F_2(a) = A \\ c_1 G_1(b) + c_2 G_2(b) = B \end{cases}$

where

$$hy_i(x) + h'y_i'(x) = F_i(x), \qquad ky_i(x) + k'y_i'(x) = G_i(x), \qquad (i = 1, 2)$$

For this, it is necessary and sufficient that the determinant of the coefficients

(5) $\qquad\qquad\qquad \Delta = \begin{vmatrix} F_1(a) & F_2(a) \\ G_1(b) & G_2(b) \end{vmatrix}$

be *non-zero*.

If, on the other hand, $A = B = 0$, i.e. *if the boundary conditions are the homogeneous conditions*

(6) $\qquad\qquad hy(a) + h'y'(a) = ky(b) + k'y'(b) = 0$

---

* See, for example, F. TRICOMI (79), Part II, pp. 327–30.
** The conditions (4) are not the most general possible, even of linear type. For example, the conditions might take the form of two given linear combinations of $y(a)$, $y'(a)$, $y(b)$, $y'(b)$ equal to A and to B. This more general case can be dealt with without difficulty by the method of integral equations treated in § 25. For further information see F. TRICOMI (82).

An example of supplementary conditions not included in (4) are the *periodicity conditions*:

$$y(a) = y(b), \qquad y'(a) = y'(b).$$

then the vanishing of the determinant Δ is a necessary and sufficient condition that the given equation possesses solutions (other than the trivial solution $y \equiv 0$) satisfying the stated boundary conditions.

Unfortunately, in practical problems, it rarely happens that two linearly independent solutions y_1 and y_2 are known, and the above discussion hinges on a knowledge of two such solutions.

However these simple considerations establish the important theorem:

THEOREM. *The necessary and sufficient condition that the differential equation* (1) *together with the arbitrary boundary conditions* (4) *possesses a solution (which is unique) is that the corresponding homogeneous problem, i.e. that for which $A = B = 0$, has no solution other than the trivial solution $y \equiv 0$.*

If the homogeneous problem admits no other solution than $y \equiv 0$, then $\Delta \neq 0$, and conversely. In other words, as the existence of two distinct solutions y and y^* of the non-homogeneous problem implies the existence of the solution $y - y^*$, which is not identically zero, of the homogeneous problem, and conversely, we may assert that *if the uniqueness theorem holds for equation* (1) *together with the boundary conditions* (4) *so also does the existence theorem, and conversely.*

Further, by considerations no less simple than those used above we may prove that

(I) *If $y_1(x)$ is an integral of* (1) *which vanishes at a point x_0 at which the leading coefficient of the equation does not vanish,*[*] *every other integral of the equation which vanishes at the point x_0 is given by the formula $y = cy_1$ where c denotes an arbitrary constant.*

If $y_2(x)$ is a second integral which vanishes at x_0, the Wronskian

$$W(x) = \begin{vmatrix} y_1(x) & y_2(x) \\ y_1'(x) & y_2'(x) \end{vmatrix}$$

of y_1 and y_2 obviously vanishes for $x = x_0$, hence is identically zero, and therefore $y_2 = cy_1$[**] where c is a suitable constant.

(II) *When the determinant Δ is zero equation* (1) *together with the homogeneous conditions* (6) *has ∞^1 solutions; in fact, if y_1 be one of these solutions all others are given by the formula $y = cy_1$ where c is an arbitrary constant.*

If y_1 and y_2 are two particular integrals of the equation which satisfy (6)

---

[*] This condition will be assumed from now onwards.
[**] See, for example, F. TRICOMI (79), Part II, p. 286.

we deduce from the first of equations (6) (the second equation need not be considered) that the two linear homogeneous equations

$$\xi y_1(a)+\eta y_1'(a) = 0, \qquad \xi y_2(a)+\eta y_2'(a) = 0$$

possess at least the one non-trivial solution $\xi = h$, $\eta = h'$. Therefore the determinant of their coefficients which coincides with $W(a)$ must be zero, and this implies that $W(x) \equiv 0$, i.e. $y_2 = cy_1$.

(III) *The zeros of the integrals of an equation of type* (1) *are all simple.**

If $y' = 0$ at a point x_0 at which $y = 0$, it follows from the uniqueness theorem of Chapter I that the integral $y(x)$ must be identically zero, as the function $y \equiv 0$ satisfies the equation and the initial conditions $y(x_0) = y'(x_0) = 0$.

17. A theorem of de la Vallée Poussin

We suppose that equation (1) is written in the *reduced form*

(7) $$y'' + p_1(x)y' + p_2(x)y = 0$$

obtained by dividing throughout by the leading coefficient.

On the assumption that the coefficients $p_1(x)$ and $p_2(x)$ are continuous, the fundamental theorem asserts that no integral $y(x)$ of the equation can have a double zero (or a zero of higher multiplicity), and consequently the interval h between two successive zeros of $y(x)$ can never be zero.

We now show that a positive lower bound can be found for values of the distance h; the theorem, originally due to de la Vallée Poussin and recently improved by Hartman and Wintner**, is as follows:

If the coefficients in equation (7) *satisfy the conditions*

(8) $$|p_1(x)| \leq M_1, \qquad |p_2(x)| \leq M_2$$

in the interval under consideration, the distance h between any two successive zeros of any integral of (7) *satisfies the inequality*

(9) $$h > \frac{\sqrt{(9M_1^2 + 24M_2)} - 3M_1}{2M_2}$$

* It is assumed in the statement of theorems of this kind that the integrals referred to are not identically zero.

** P. HARTMAN and A. WINTNER: 'On an oscillation criterion of de la Vallée Poussin,' *Quart. Applied Math.* **13**, 330–32 (1955).

provided that $M_2 > 0$. If $M_2 = 0$, then

(9') $$h > \frac{2}{M_1}$$

We begin the proof by noting that for any given function $\phi(x)$ with continuous derivative in the interval $(0, h)$

(10) $$h\phi(x) = \int_0^x \xi\phi'(\xi)\,d\xi - \int_x^h (h-\xi)\phi'(\xi)\,d\xi + \int_0^h \phi(\xi)\,d\xi$$

Since on integrating by parts we have

$$\int_0^x \xi\phi'(\xi)\,d\xi = x\phi(x) - \int_0^x \phi(\xi)\,d\xi$$

and

$$\int_x^h (h-\xi)\phi'(\xi)\,d\xi = -(h-x)\phi(x) + \int_x^h \phi(\xi)\,d\xi$$

the identity (10) follows immediately on subtraction term by term.

Now suppose that some integral $y(x)$ of (7) has two successive zeros at $x = 0$ and $x = h$,* and apply the identity (10) to the function $\phi(x) \equiv y'(x)$, noting that

$$\int_0^h \phi(\xi)\,d\xi = \int_0^h y'(\xi)\,d\xi = y(h) - y(0) = 0$$

so that

$$hy'(x) = \int_0^x \xi y''(\xi)\,d\xi - \int_x^h (h-\xi)y''(\xi)\,d\xi$$

On substitution for y'' its value obtained from (7) this yields

(11) $$hy'(x) = -\int_0^x \xi p_1(\xi)y'(\xi)\,d\xi + \int_x^h (h-\xi)p_1(\xi)y'(\xi)\,d\xi$$
$$- \int_0^x \xi p_2(\xi)y(\xi)\,d\xi + \int_x^h (h-\xi)p_2(\xi)y(\xi)\,d\xi$$

Now let μ denote the maximum (certainly positive)** of $|y'(x)|$ in $(0, h)$; consequently, since $y(x)$ vanishes for both $x = 0$ and $x = h$, both inequalities

$$|y(\xi)| \leq \mu\xi, \qquad |y(\xi)| \leq \mu(h-\xi)$$

---

* There is no restriction in supposing one of the two zeros to be at the origin since (7) is unaltered by the transformation $x = \xi + a$.

** It is obvious that if $\mu = 0$ then, since $y'(x) \equiv 0$, the integral $y(x)$ must be the trivial integral $y(x) \equiv 0$ which is excluded here.

are satisfied in the interval $(0, h)$, so that neither product

$$\xi y(\xi) \quad \text{and} \quad (h-\xi) y(\xi)$$

exceeds in absolute value the quantity

$$\mu\xi(h-\xi)$$

Employing the inequalities (8), we deduce from (11) that

$$h|y'(x)| < M_1\mu\left[\int_0^x \xi\, d\xi + \int_x^h (h-\xi)\, d\xi\right] + M_2\mu\int_0^h \xi(h-\xi)\, d\xi$$

from which follows

(12) $$|y'(x)| < M_1\mu\frac{h}{2} + M_2\mu\frac{h^2}{6} \qquad (0 \le x \le h)$$

since

$$\int_0^h \xi(h-\xi)\, d\xi = \frac{h^3}{6}$$

and

$$\int_0^x \xi\, d\xi + \int_x^h (h-\xi)\, d\xi = \tfrac{1}{2}[x^2 + (h-x)^2]$$

whose minimum value $h^2/4$ occurs when $x = h/2$ and maximum value $h^2/2$ occurs when $x = 0$ and $x = h$, so that this expression is always not greater than $h^2/2$.

By applying (12) at a point at which $|y'(x)| = \mu$ (which point certainly exists, by Weierstrass' theorem) and dividing throughout by μ, which is not zero, we obtain

$$1 < M_1\frac{h}{2} + M_2\frac{h^2}{6}$$

i.e. h must satisfy the inequality

$$\tfrac{1}{6}M_2 h^2 + \tfrac{1}{2}M_1 h - 1 > 0$$

Hence h must lie outside the interval whose extremes are the roots of the equation

$$\tfrac{1}{6}M_2 h^2 + \tfrac{1}{2}M_1 h - 1 = 0$$

viz.

$$-\frac{\sqrt{(9M_1^2 + 24M_2)} + 3M_1}{2M_2}, \quad \frac{\sqrt{(9M_1^2 + 24M_2)} - 3M_1}{2M_2}$$

and the inequality (9) follows. If $M_2 = 0$, then

$$\tfrac{1}{2}M_1 h - 1 > 0$$

and the inequality (9') follows.

It is instructive to apply this theorem to the equation of the harmonic oscillator

(13) $$y'' + k^2 y = 0$$

(where k^2 denotes a given constant) whose general integral may be written as

$$y = A \sin k(x - \gamma)$$

where A and γ are arbitrary constants. The zeros of any integral of (13) are therefore uniformly distributed on the x-axis at a distance of π/k from one another. Since in this case (within any interval) $M_1 = 0$, $M_2 = k^2$, the inequality (9) gives $h > \sqrt{6}/k$ and consequently $\pi/k > \sqrt{6}/k$, i.e.

$$\pi > \sqrt{6}$$

The theorem therefore supplies an exact lower bound (although not a close one) for π.

It follows from the theorem of de la Vallée Poussin that *the zeros of any integral of an equation of the type* (7) *cannot possess a point of accumulation in any finite interval in which the coefficients $p_1(x)$ and $p_2(x)$ remain bounded.** In fact the greatest number of zeros which can fall within an interval of length l in which the inequalities (8) are satisfied cannot exceed by more than one the integral part of l/r, where r is the positive root of the equation

$$\tfrac{1}{6} M_2 h^2 + \tfrac{1}{2} M_1 h - 1 = 0$$

In other words—if we denote by the *conjugates* of a point x_0 the further zeros of the integrals of (7) which vanish for $x = x_0$,** the preceding theorem asserts that *the conjugates of a certain point x_0 form a discrete set*;*** this implies that it is permissible to refer to a *first* conjugate to the right or to the left of x_0, to a *second* conjugate, etc.

But do such conjugates actually exist?

In some cases, as for example in (13), the answer is certainly yes. But in other cases, as for example for the equation

(14) $$y'' - k^2 y = 0$$

they may be non-existent; for the general integral of (14) may be written as

$$y = A \sinh k(x - \gamma)$$

---

* This result might otherwise be deduced from the more elementary theorem III of § 16.

** To determine any one of these integrals we use the initial conditions $y(x_0) = 0$, $y'(x_0) = y'_0$ where y'_0 is any non-zero constant.

*** A set of points is called *discrete* if it possesses no finite points of accumulation.

which shows that all the integrals vanishing for $x = x_0$, i.e. the ∞^1 integrals corresponding to the value x_0 of the constant γ, vanish for no $x \neq x_0$ since the *hyperbolic sine* vanishes only at the origin.

With regard to the integrals of an equation of the type (7) there may therefore arise two distinct possibilities whose importance will be seen more clearly later:

(1) either such integrals are *oscillatory*, i.e. they vanish in at least two points in the interval under consideration, and consequently every point x_0 possesses at least one conjugate;

or

(2) the integrals are *not oscillatory*, i.e. they vanish at most once within the interval under consideration and there then exists no pair of conjugate points.

18. Simplifications of the given equation

Before discussing equations of type (1) further, we note that if $B(x) \equiv A'(x)$ equation (1) may be written in the particular form

$$\frac{d}{dx}[A(x)y'] + C(x)y = 0$$

This form is called *self-adjoint* and is extremely useful. In fact, any given equation (1) may be written in this form, by multiplying (1) throughout by a suitable factor $H(x)$ obtained by a quadrature. This is seen by considering the function $H(x)$ determined by the equation

$$\frac{d}{dx}[A(x)H(x)] = B(x)H(x)$$

from which, on expansion and division throughout by $A(x)H(x)$, we derive

$$\frac{H'(x)}{H(x)} = -\frac{A'(x)}{A(x)} + \frac{B(x)}{A(x)}$$

and on integration

$$\log H(x) = -\log A(x) + \int \frac{B(x)}{A(x)} dx$$

so that

$$H(x) = \frac{1}{A(x)} e^{\int \frac{B(x)}{A(x)} dx}$$

Now if we put

(15) $$p(x) = e^{\int \frac{B(x)}{A(x)} dx}, \qquad P(x) = \frac{C(x)}{A(x)} e^{\int \frac{B(x)}{A(x)} dx}$$

the original equation may be written in the form

(16) $$\frac{d}{dx}\left[p(x)\frac{dy}{dx}\right] + P(x)y = 0$$

Under the usual conditions that $A(x)$, $B(x)$, $C(x)$ be continuous and also that $A(x) \neq 0$, the function $p(x)$ in addition to being continuous (as is also P) is always positive with continuous first derivative derived from the first equation in (15),

$$p'(x) = \frac{B(x)}{A(x)} e^{\int \frac{B(x)}{A(x)} dx}$$

By now making a change of independent variable x to the new variable ξ defined by the relation

(17) $$\xi = \int \frac{dx}{p(x)}$$

equation (16) may be further reduced to *the case in which $p(x) \equiv 1$*.
For from (17) we have

$$\frac{d\xi}{dx} = \frac{1}{p(x)}, \qquad \frac{d}{dx} = \frac{1}{p(x)}\frac{d}{d\xi}$$

and therefore (16) takes the form

$$\frac{1}{p(x)}\frac{d}{d\xi}\left[\frac{dy}{d\xi}\right] + P(x)y = 0$$

i.e.

(18) $$\frac{d^2 y}{d\xi^2} + Q(\xi)y = 0$$

where

(19) $$Q(\xi) = p(x)\,P(x)$$

When convenient therefore we shall refer to equation (18) which contains a single coefficient; for other purposes it will appear that (16) is the more suitable form leading sometimes to as simple results as those deducible from (18).* Further transformations will be discussed in § 32.

* Note that if two or more equations of the form (16) involving different functions $p(x)$ are considered it is not possible to reduce them to the form (18) by the *same* transformation (17).

19. Theorems on the zeros and on the maxima and minima of integrals

By easy applications of the preceding transformations we shall prove the following theorem which is of importance although its proof is very simply obtained.

No integral of (16) *in which* $p(x) > 0$ *can oscillate in any interval at all points of which* $P(x) \leq 0$.*

Let
$$v(x) = p(x)\, y(x)\, y'(x)$$
then, by (16),
$$v'(x) = (py')'y + py'^2 = py'^2 - Py^2$$

Hence in any interval in which $P \leq 0$ we have $v'(x) \geq 0$. But this implies that if the function $v(x)$ vanishes in this interval it can vanish no more than once; hence, as $v(x)$ vanishes when $y(x)$ vanishes, no integral $y(x)$ of (16) can vanish more than once in any interval throughout which $P(x) \leq 0$.

This proves further that y' cannot vanish more than once, and that if y vanishes then y' *cannot* vanish, and conversely.**

With reference to (18) instead of (16), the theorem just proved assumes an intuitive geometrical character—for in any interval throughout which $P \leq 0$ (and consequently $Q \leq 0$), by virtue of (18),

$$y\frac{d^2y}{d\xi^2} = -Q(\xi)y^2 \geq 0$$

which implies that the integral curves continually turn away from the ξ-axis, and this property is clearly inconsistent with the existence of more than one point of intersection with the ξ-axis.

By methods similar to those employed in the last theorem, i.e. by constructing a suitable auxiliary monotonic function similar to the function $v(x)$, it is fairly easy to establish the following elegant theorem of Sonin-Pólya***:

---

* The case $P(x) > 0$ will be discussed later in § 21 under the assumption that $p(x) \equiv 1$.

** Cf. M. Picone and C. Miranda: *Esercizi di Analisi Matematica*, p. 442 (2nd edition, Rome, Studium Urbis, 1945). If instead of referring to the self-adjoint equation (16) we refer to the original equation (1) the condition $P(x) \leq 0$ must be replaced by $C(x)/A(x) \leq 0$, as is immediately seen from the second relation (15) (assuming that $A(x) \neq 0$).

*** This theorem proved initially by Sonin for the case $p(x) \equiv 1$ has recently been generalized by Pólya to cover equations of the form (16) (cf. G. Szegö (78), p. 161) in which $p(x) > 0$, $P(x) > 0$, conditions which in this case are fulfilled provided that $P \neq 0$. It follows from the preceding theorem that if $p(x) P(x) < 0$ no integral can be oscillatory.

§ 19]   *Theorems on the zeros and on the maxima and minima of integrals*   99

If the coefficients $p(x)$ and $P(x)$ which appear in (16) together with their first derivatives are continuous in a certain interval (a, b), and if the product $p(x)P(x)$ is a non-decreasing (non-increasing) function of x in that interval and $P(x)$ never vanishes there, the possible maxima and minima occurring in (a, b) of any integral $y(x)$ of (16) are such that the corresponding values of $|y|$ form a non-increasing (non-decreasing) sequence.

Consider the function

$$\phi(x) = y^2(x) + \frac{p(x)}{P(x)} y'^2(x) = y^2 + \frac{1}{pP}(py')^2$$

with derivative

$$\phi'(x) = 2yy' + \frac{2py'}{pP}(py')' - \frac{(pP)'}{(pP)^2}(py')^2$$

By (16) which states that

$$(py')' = -Py$$

this may be written in the form

$$\phi'(x) = -\left[\frac{y'(x)}{P(x)}\right]^2 \frac{d}{dx}[p(x)P(x)]$$

showing that within (a, b) the derivative $\phi'(x)$ is non-positive if the derivative of $p(x)P(x)$ is non-negative, and vice versa. But at a maximum or minimum point of y where $y' = 0$, $\phi = y^2$; hence the squares of the maxima and minima of y, and hence the corresponding values of $|y|$, form a *non-increasing* sequence when $p(x)P(x)$ is *non-decreasing*, and vice versa.*

It follows as a corollary from this theorem that in the part of the interval (a, b) in which $\phi(x) \leq 0$ there can occur no maxima nor minima of y, and that by considering the greatest of the three numbers $|y(a)|$, $|y(b)|$ and $\sqrt{\Phi}$ where Φ denotes the greatest value (supposed positive) of $\phi(x)$ in (a, b), we can derive an upper bound for the values of $|y(x)|$ within the interval itself.

Further, we note that the equality $\phi = y^2$ holds not only at points at which $y' = 0$ but also at any point at which $p(x) = 0$. Consequently the values of $|y(x)|$ at such points may be considered as included in the sequence which has just been proved monotonic.

In the proofs of the two preceding theorems an essential part was played by two particular quadratic forms in y and y' (in the first case the form pyy' and in the second case the form ϕ) which under certain conditions were

---

* If we use the original form (1) of the equation, in place of (16), the corresponding condition is

$$\operatorname{sgn} \phi'(x) = -\operatorname{sgn}[C'(x) A(x) - C(x) A'(x) + 2B(x) C(x)]$$

This is easily deduced from (15), assuming throughout that $A(x) \neq 0$.

shown to be monotonic. The possibility is therefore suggested of obtaining other theorems of the same type by considering in place of the preceding forms other suitable quadratic forms in y and y', and this has been carried out effectively by the author's former assistant U. Richard.* Among numerous results which may be thus obtained one of the simplest is that derived from the quadratic form

$$\psi(x) = p(x)P(x)\phi(x) = p(x)P(x)y^2(x) + p^2(x)y'^2(x)$$

with derivative

$$\psi'(x) = \phi(x)\frac{d}{dx}[p(x)P(x)] + \phi'(x)p(x)P(x) = y^2(x)\frac{d}{dx}[p(x)P(x)]$$

This yields the result that *under the same conditions as those stated in the theorem of Sonin-Pólya, the values of*

$$|p(x)P(x)|^{\frac{1}{2}}|y(x)|$$

*calculated at the points in which $y'(x)$ or $p(x) = 0$ form a monotonic sequence of the same character as that formed by the values of $p(x)P(x)$.***

In the case in which an integral $y(x)$ of (16) has *infinite* maxima and minima at $x_1, x_2, \ldots$, some simple results have been obtained regarding the asymptotic character of the sequence $|y(x_1)|, |y(x_2)|, \ldots$. In particular, U. Richard,*** by generalizing previous results of H. Milloux and A. Wiman,**** has recently proved the following theorem, for the proof of which (not difficult but not short) the reader is referred to the original paper.

If for all x to the right of a certain point $x = a$, $p(x) > 0$, $P(x) > 0$, and if the function

$$\omega(x) = \tfrac{1}{2}p(x)\frac{d}{dx}\{[p(x)P(x)]^{-\frac{1}{2}}\}$$

tends monotonically to zero as $x \to +\infty$, with every integral $y(x)$ of (16) may be associated a certain positive constant k such that, if $x_1, x_2, \ldots$, be successive maxima and minima points of this integral, then

$$[(pP)^{\frac{1}{2}}|y|]_{x=x_n} = k + O[\omega(x_n)] \qquad (n \to +\infty)$$

---

* U. Richard: 'Sulle successioni di valori stazionari delle soluzioni di equazioni differenziali lineari del secondo ordine,' *Rend. Seminario Mat. Torino*, **9**, 309–24 (1949–50); also 'Alcuni problemi asintotici per le equazioni differenziali lineari,' *Rend. Seminario Mat. Torino*, **15**, 59–64 (1955–56).
** That is, *increasing* if the values of pP are increasing, and *decreasing* if the values of pP are decreasing.
*** U. Richard: *Rend. Lincei*, (8) **12**, 382–87 (1952).
**** H. Milloux: 'Sur l'équation différentielle $x'' + A(t)x = 0$,' *Prace Mat. Fiz.*, **41**, 39–54 (1934); A. Wiman: 'Über eine Stabilitätsfrage in der Theorie der linearen Differentialgleichungen,' *Acta Math.*, **66**, 121–45 (1936).

20. Comparison theorems and their corollaries

There is a celebrated theorem on the zeros of an equation of type (16) due to Sturm, a pioneer in the *direct* study of differential equations, which in many cases allows us to derive important results by *comparing* the given equation with another of the same type which can be explicitly integrated or whose zeros are known. We consider, for simplicity, two equations involving the same function $p(x)$*.

THE COMPARISON THEOREM OF STURM. *Given the two differential equations*

$$(20) \quad \frac{d}{dx}\left[p(x)\frac{dy_1}{dx}\right]+P_1(x)y_1 = 0, \quad \frac{d}{dx}\left[p(x)\frac{dy_2}{dx}\right]+P_2(x)y_2 = 0$$

in which $P_2(x) \geq P_1(x)$; then between two successive zeros of an integral of the first equation there necessarily lies at least one zero of any integral of the second equation.

We shall establish the theorem by showing the contrary, viz. that if a and b are two successive zeros of a certain integral $\bar{y}_1(x)$ of the first equation (20) and if $P_2(x) \geq P_1(x)$ for $a \leq x \leq b$, there exists an integral $\bar{y}_2(x)$ of the second which does not vanish in $a \leq x \leq b$, to be absurd.

As the two integrals $\bar{y}_1$ and $\bar{y}_2$ introduced above may be multiplied by a constant factor without altering their zeros, the properties in which we are interested here, we may suppose that in (a, b) we have $\bar{y}_1(x) \geq 0$, $\bar{y}_2(x) > 0$. This implies that of the two derivatives $\bar{y}'_1(a)$ and $\bar{y}'_1(b)$ (which are certainly not null as a and b are not multiple zeros of $\bar{y}_1$) the first must be *positive* and the second *negative*, i.e.

$$(21) \quad \bar{y}'_1(a)>0, \quad \bar{y}'_1(b)<0$$

since these are respectively derived from the ratios

$$\frac{\bar{y}_1(x)-\bar{y}_1(a)}{x-a}, \quad \frac{\bar{y}_1(x)-\bar{y}_1(b)}{x-b}$$

* If in place of the transformation (17) we use the transformation

$$\int \frac{d\xi}{p_1(\xi)} = \int \frac{dx}{p(x)}$$

equation (16) is replaced by a similar equation in which the function $p(x)$ is replaced by $p_1(\xi)$.

and for x within the interval (a, b),

$$\bar{y}_1(x) - \bar{y}_1(a) = \bar{y}_1(x) > 0, \quad x - a > 0$$
$$\bar{y}_1(x) - \bar{y}_1(b) = \bar{y}_1(x) > 0, \quad x - b < 0$$

Now multiplying the equations (20) respectively by y_2 and y_1 and subtracting one equation from the other, we obtain the identity

$$(22) \quad [P_2(x) - P_1(x)] y_1 y_2 = y_2 \frac{d}{dx}\left[p(x)\frac{dy_1}{dx}\right] - y_1 \frac{d}{dx}\left[p(x)\frac{dy_2}{dx}\right]$$

$$= \frac{d}{dx}\left[p(x)\left(y_2 \frac{dy_1}{dx} - y_1 \frac{dy_2}{dx}\right)\right]$$

from which, by replacing y_1 and y_2 by $\bar{y}_1$ and $\bar{y}_2$ respectively and integrating between a and b, we derive

$$\int_a^b [P_2(x) - P_1(x)] \bar{y}_1(x) \bar{y}_2(x) dx = \left[p(x)\left(\bar{y}_2 \frac{d\bar{y}_1}{dx} - \bar{y}_1 \frac{d\bar{y}_2}{dx}\right)\right]_a^b$$

i.e.

$$(23) \quad \int_a^b [P_2(x) - P_1(x)] \bar{y}_1(x) \bar{y}_2(x) dx = p(b)\bar{y}_2(b)\bar{y}_1'(b) - p(a)\bar{y}_2(a)\bar{y}_1'(a)$$

But this equality is *absurd* since the left-hand side is non-negative in view of the conditions $P_2(x) \geq P_1(x)$, $\bar{y}_1(x) \geq 0$, $\bar{y}_2(x) > 0$, and the right-hand side is certainly negative on account of (21). Hence the function $\bar{y}_2(x)$ vanishes at least once within the interval (a, b) or at an end-point.*

As a particular case the theorem of Sturm may be applied to the equation itself, giving the following corollary:

Between two successive zeros of any integral of a linear differential equation of the second order there lies exactly one zero of every other integral linearly independent of the first.

If the integral $y_1(x)$ of the equation has two successive zeros at the points $x = a$ and $x = b$, every other integral $y_2(x)$ of the equation either vanishes itself for $x = a$ so that it can differ only by a constant factor from $y_1(x)$; or else it vanishes within the interval (a, b) in which case it can vanish only once, for if it vanished two (or more) times, then between its successive zeros must fall at least one zero of $y_1(x)$, contrary to the hypothesis that a and b are successive zeros of this integral.

* It should be noted that if $y_2(x)$ does not vanish *within* the interval (a, b) it must vanish at *both* extremes of the interval, and therefore $P_1(x) \equiv P_2(x)$, for otherwise (23) would be absurd.

Comparison theorems and their corollaries

This corollary may be more concisely stated as—*the zeros of two linearly independent integrals of a linear differential equation of the second order interlace each other.* (The separation theorem.)

Before proceeding to further applications of Sturm's theorem we prove the following comparison theorem which will also be used later:

THEOREM OF NUMERICAL COMPARISON. *If $P_1(x) < P_2(x)$ and if $\bar{y}_1(x)$ and $\bar{y}_2(x)$ are two integrals of the equations* (20) *satisfying the same initial conditions*

(24) $$\bar{y}_1(x_0) = \bar{y}_2(x_0) = y_0, \qquad \bar{y}'_1(x_0) = \bar{y}'_2(x_0) = \bar{y}'_0$$

then in any neighbourhood to the right of the point x_0 in which $\bar{y}_1(x)$ and $\bar{y}_2(x)$ are never zero (except possibly at $x = x_0$),

(25) $$|\bar{y}_1(x)| > |\bar{y}_2(x)|$$

Further, the ratio $\bar{y}_1(x)/\bar{y}_2(x)$ is a function which increases with x from the value 1 *assumed for $x = x_0$.*

From the identity (22) above, on integrating from x_0 to $x > x_0$ and noting that

$$\left[p(x)\left(\bar{y}_2 \frac{d\bar{y}_1}{dx} - \bar{y}_1 \frac{d\bar{y}_2}{dx} \right) \right]_{x=x_0} = p(x_0)(y_0 y'_0 - y_0 y'_0) = 0$$

we have

(26) $$p(x)\left(\bar{y}_2 \frac{d\bar{y}_1}{dx} - \bar{y}_1 \frac{d\bar{y}_2}{dx} \right) = \int_{x_0}^{x} [P_2(x) - P_1(x)] \bar{y}_1(x) \bar{y}_2(x) dx$$

Now the integral on the right is certainly positive, as $P_2(x) > P_1(x)$ and the continuous function $\bar{y}_1(x)\bar{y}_2(x)$ (which cannot vanish and therefore cannot change sign between x_0 and x) has the same sign as it has within the immediate neighbourhood of x_0, i.e. is positive, since if $y_0 \neq 0$ then $\bar{y}_1(x_0)\bar{y}_2(x_0) = y_0^2 > 0$, while if $y_0 = 0$ (and then y'_0 cannot be zero) we have

$$\lim_{x \to x_0} \frac{\bar{y}_1(x)\bar{y}_2(x)}{(x-x_0)^2} = \lim_{x \to x_0} \frac{\bar{y}_1(x) - \bar{y}_1(x_0)}{x - x_0} \cdot \lim_{x \to x_0} \frac{\bar{y}_2(x) - \bar{y}_2(x_0)}{x - x_0} = y_0'^2$$

It follows therefore, since $p(x) > 0$, that to the right of the point x_0

(27) $$\bar{y}_2 \frac{d\bar{y}_1}{dx} - \bar{y}_1 \frac{d\bar{y}_2}{dx} > 0$$

i.e. $$\bar{y}_2^2 \frac{d}{dx}\left(\frac{\bar{y}_1}{\bar{y}_2} \right) > 0$$

which implies that $\bar{y}_1(x)/\bar{y}_2(x)$ is an increasing function of x, and, in addition, that

$$\frac{\bar{y}_1(x)}{\bar{y}_2(x)} > 1, \quad \text{for } x > x_0$$

The result (25) follows immediately. Even in the case $y_0 = 0$, $y'_0 \neq 0$, the limit of $\bar{y}_1(x)/\bar{y}_2(x)$ as $x \to x_0$ is 1; for, by l'Hospital's rule,

$$\lim_{x \to x_0} \frac{\bar{y}_1(x)}{\bar{y}_2(x)} = \lim_{x \to x_0} \frac{\bar{y}'_1(x)}{\bar{y}'_2(x)} = \frac{y'_0}{y'_0} = 1$$

Finally it should be noted that if in place of the inequality $P_2(x) > P_1(x)$ we have the *weaker* inequality $P_2(x) \geqq P_1(x)$, and provided that the two continuous functions $P_1(x)$ and $P_2(x)$ are not always equal (in which case $\bar{y}_1(x)$ and $\bar{y}_2(x)$ coincide), it can likewise be asserted that (25) and (27) hold, as the integral on the right of (26) is still necessarily positive.

21. The interval between successive zeros of an integral

From the comparison theorem of Sturm may be obtained a double inequality for the distance between two successive zeros of an integral of a linear equation of second order; this inequality greatly improves on the one bound established by the theorem of de la Vallée Poussin.

Here we suppose the equation reduced to the form (18)

$$\frac{d^2 y}{dx^2} + Q(x)y = 0$$

and we suppose that in the interval considered *the coefficient Q is always positive*, i.e. that there exist two positive numbers m and M where $M \geqq m$ such that

(28) $$m^2 \leqq Q(x) \leqq M^2$$

If δ be the distance between two successive zeros of any integral whatsoever of the equation, then

(29) $$\frac{\pi}{M} \leqq \delta \leqq \frac{\pi}{m}$$

In view of (28) we can use the theorem of Sturm to compare the given equation with the two equations with constant coefficients

$$\text{(30)} \qquad \frac{d^2 y}{dx^2} + m^2 y = 0 \quad \text{and} \quad \frac{d^2 y}{dx^2} + M^2 y = 0$$

whose general integrals are respectively

$$y = A \sin m(x - \gamma), \qquad y = A^* \sin M(x - \gamma^*)$$

Hence the conjugates of a point x_0 with respect to the first equation (30) are the points $x_0 + n\pi/m$, where n is integral, and with respect to the second equation (30) the points $x_0 + n\pi/M$. The theorem of Sturm asserts that if $\bar{y}(x)$ be any integral whatsoever of the given equation then between two successive zeros x_0 and x_1 of $\bar{y}(x)$ must fall at least one zero of any integral whatsoever of the second equation (30); hence this interval must contain the first conjugate to the right, $x_0 + \pi/M$, of the point x_0*; similarly between x_0 and $x_0 + \pi/m$ which are two successive zeros of an integral of the first equation (30) there must fall at least one zero of $\bar{y}(x)$ distinct from x_0; hence x_1 falls in this interval; hence follow the two inequalities

$$x_0 + \frac{\pi}{M} \leq x_1, \qquad x_1 \leq x_0 + \frac{\pi}{m}$$

and on putting $x_1 - x_0 = \delta$ these become (29).

In the theorem of de la Vallée Poussin we can put $M_1 = 0$, $M_2 = M^2$, $h = \sqrt{6}/M$, which gives, in place of (29), as lower limit $\delta > \sqrt{6}/M$.**

One corollary of the last theorem is that *if* (28) *is satisfied throughout an interval* (α, β) *of length not less than* π/m, *then all the integrals of the given equation must be oscillatory in that interval.* For $\delta < \pi/m$ and therefore within the interval (α, β) there must fall at least *two* zeros of each integral of the given equation.

Further, if we denote by N_h the number of zeros of an integral of the given equation which lie in an (open) interval of length h in which (28) (or even only the single inequality $Q(x) \geq m^2$) is satisfied, then

$$\text{(31)} \qquad N_h \geq \left[\frac{hm}{\pi}\right] - 1$$

---

* Since $x_0 + \pi/M$ is the first zero to the right of x_0 of those integrals of the second equation (30) which vanish for $x = x_0$, i.e. of the integrals of the form $y = A^* \sin M(x - \gamma^*)$ when $\gamma^* = x_0$.

** The inequalities (29) cannot be improved without introducing more restrictive conditions on the coefficient $Q(x)$ than those of (28) since the bounds derived from these are in fact attained when $Q(x)$ is a constant.

where the square brackets denote the integral part of the number contained within them. In fact the N_h zeros and the two extremes of the interval (whether or not they themselves are zeros) determine $N_h + 1$ intervals each of which has length not exceeding π/m, and the sum of their lengths is h. Therefore $(N_h + 1)\pi/m \geq h$, from which follows (31).

As example we apply (29) to the *Bessel equation*

$$(32) \qquad \frac{d}{dx}\left(x\frac{dy}{dx}\right)+\left(x-\frac{v^2}{x}\right)y = 0$$

where v is a constant (which is here assumed positive). This equation which will be discussed in more detail later is one of the most important in analysis. It may be reduced by making the transformation

$$y\sqrt{x} = z$$

more easily than by the general method of reduction, and the reduced equation is

$$(32') \qquad \frac{d^2z}{dx^2}+\left(1+\frac{\frac{1}{4}-v^2}{x^2}\right)z = 0$$

In this case therefore

$$Q(x) = 1+\frac{\frac{1}{4}-v^2}{x^2}$$

and we must evidently exclude values of x in the vicinity of the origin at which the coefficient $Q(x)$ becomes infinite. It is therefore convenient to consider values of x not less than a positive quantity η which may be whatever we please if $v^2 \leq 1/4$, but if $v^2 > 1/4$ we take

$$\eta > \sqrt{(v^2 - \tfrac{1}{4})}$$

Under these conditions we may assert that

$$\begin{cases} 1 \leq Q(x) \leq 1+\dfrac{\frac{1}{4}-v^2}{\eta^2} & \text{(if } v^2 \leq \tfrac{1}{4}) \\[2mm] 0 < 1-\dfrac{v^2-\frac{1}{4}}{\eta^2} \leq Q(x) \leq 1 & \text{(if } v^2 \geq \tfrac{1}{4}) \end{cases}$$

This implies, by (29), that for $x \geq \eta$ the distance δ between two successive zeros of a *Bessel function** satisfies the inequalities

(33)
$$\begin{cases} \pi\left(1+\dfrac{\frac{1}{4}-v^2}{\eta^2}\right)^{-\frac{1}{2}} \leq \delta \leq \pi & (v^2 \leq \tfrac{1}{4}) \\[2ex] \pi \leq \delta \leq \pi\left(1-\dfrac{v^2-\frac{1}{4}}{\eta^2}\right)^{-\frac{1}{2}} & (v^2 \geq \tfrac{1}{4}) \end{cases}$$

In particular, for $v^2 = 1/4$ we find $\delta = \pi$, in accordance with the fact that the general integral of the Bessel equation is then (see (32'))

$$C_1 \frac{\sin x}{\sqrt{x}} + C_2 \frac{\cos x}{\sqrt{x}} \text{**}$$

Having now obtained a non-null lower bound for δ we can assert that in all cases (provided that $x \geq \eta$) *every solution of the Bessel equation is ultimately oscillatory and possesses an infinity of zeros the distance between any consecutive two of which tends to π as x increases indefinitely.*

This is so since the left side of the first inequality in (33) and the right side of the second inequality in (33) both tend to π as $\eta \to \infty$.

22. An important change of variable

Prüfer*** has suggested comparatively recently an important change of variable which in certain cases has proved extremely useful. This consists in replacing the differential equation (16) by the equivalent normal system

(34)
$$\frac{dy}{dx} = \frac{1}{p(x)} z, \qquad \frac{dz}{dx} = -P(x)y$$

and then replacing y and z by *polar coordinates*, viz.

(35) $$y = \rho \sin \theta, \qquad z = \rho \cos \theta$$

Thus

$$\rho \frac{d\rho}{dx} = y\frac{dy}{dx} + z\frac{dz}{dx}, \qquad \rho^2 \frac{d\theta}{dx} = z\frac{dy}{dx} - y\frac{dz}{dx}$$

---

* Any solution of equation (32) is called a Bessel function.

** It should be noted that the particular integral with C_1 as multiplicative constant remains finite, and in fact tends to zero, as $x \to 0$, while the second integral tends to infinity.

*** H. PRÜFER: 'Neue Herleitung der Sturm-Liouvilleschen Reihenentwicklung stetiger Funktionen,' *Math. Ann.*, **95**, 499–518 (1926).

System (34) may therefore be replaced by

$$\begin{cases} \rho \dfrac{d\rho}{dx} = \rho^2 \left[\dfrac{1}{p(x)} - P(x) \right] \sin\theta \cos\theta \\ \rho^2 \dfrac{d\theta}{dx} = \rho^2 \left[\dfrac{1}{p(x)} \cos^2\theta + P(x) \sin^2\theta \right] \end{cases}$$

i.e.

(36)
$$\begin{cases} \dfrac{d\theta}{dx} = \dfrac{1}{p(x)} \cos^2\theta + P(x) \sin^2\theta \\ \dfrac{d\log\rho}{dx} = \dfrac{1}{2} \left[\dfrac{1}{p(x)} - P(x) \right] \sin 2\theta \end{cases}$$

in which form we lose the trivial solution $y \equiv 0$.

The system just obtained is important on account of the fact that the first equation (36) contains only the unknown function $\theta(x)$ and not $\rho(x)$, so that in substance the problem is reduced to the integration of this equation of the first order.* When the function $\theta(x)$ has been found from the first equation in (36) the function $\rho(x)$ can be immediately obtained by a simple quadrature, viz.

(37)
$$\log \frac{\rho}{\rho_a} = \frac{1}{2} \int_a^x \left[\frac{1}{p(x)} - P(x) \right] \sin 2\theta(x)\, dx$$

where ρ_a indicates the (arbitrary) value of ρ at some point $x = a$.

There is also the further advantage that of the two arbitrary constants introduced in integration only one (that arising from the integration of the first equation in (36), which might, for example, be the value θ_a of $\theta(x)$ for $x = a$) plays an essential part, while the other, the constant arising from (37), merely restates the property that if $y_0(x)$ be one solution of the problem then $Cy_0(x)$, where C is an arbitrary constant, is also a solution—provided that any supplementary conditions adjoined to equation (16) are homogeneous. This occurs, for example, for (6) of § 16.

It is therefore clear that when a *Sturm-Liouville system*, i.e. an equation of the form (16) together with the boundary conditions (6), is integrated by

---

* It is not surprising that the integration of a *linear* differential equation of second order can be reduced to that of an equation (no longer linear) of first order. It is well known (see Bieberbach (4), p. 26, for example) that equations of second order can be easily transformed into equations of first order of the Riccati type, i.e. an equation in which the derivative equals a second degree polynomial in the unknown function, and the first equation (36) can be written as the *Riccati equation*

$$\frac{dt}{dx} = \frac{1}{p(x)} + P(x) t^2$$

on making the change of variable $t = \tan\theta$.

means of the Prüfer change of variable the constant ρ_a which appears above may have any value; but this is not so for the values θ_a and θ_b assumed by the function $\theta(x)$ at the extremities of the interval (a, b). These values θ_a and θ_b must be such that

(38) $$h \sin \theta_a + \frac{h'}{p(a)} \cos \theta_a = 0, \qquad k \sin \theta_b + \frac{k'}{p(b)} \cos \theta_b = 0$$

On introducing the angles α and β such that

$$h = H \cos \alpha, \quad -\frac{h'}{p(a)} = H \sin \alpha; \quad k = K \cos \beta, \quad -\frac{k'}{p(b)} = K \sin \beta$$

where H and K denote factors of proportionality given by

(39) $$H = \pm\sqrt{(h^2 + [h'/p(a)]^2)}, \qquad K = \pm\sqrt{(k^2 + [k'/p(b)]^2)}$$

equations (38) may be written as

$$\sin(\theta_a - \alpha) = 0, \qquad \sin(\theta_b - \beta) = 0$$

i.e.

(40) $$\theta_a = \alpha + n_1 \pi, \qquad \theta_b = \beta + n_2 \pi$$

where n_1 and n_2 denote any two integers, and α and β are determined by

(41) $$\tan \alpha = -\frac{h'}{hp(a)}, \qquad \tan \beta = -\frac{k'}{kp(b)}$$

and the inequalities

(42) $$0 \leq \alpha < \pi, \qquad 0 < \beta \leq \pi$$

Thus, for example, in the case of the so-called *first boundary-value problem* (i.e. the case in which the supplementary conditions are (3) of § 16 together with $A = B = 0$, and therefore $h' = k' = 0$) we have

$$\alpha = 0, \qquad \beta = \pi$$

With the introduction of the Prüfer change of variables the solution of a Sturm-Liouville system is therefore reduced to the integration of the first equation of (36) which we rewrite here as

(43) $$\frac{d\theta}{dx} = \frac{1}{p(x)} \cos^2 \theta + P(x) \sin^2 \theta$$

together with the two conditions (40); this implies, as has already been pointed out in § 16, that in general a system of this kind has no solution apart from the trivial solution $y \equiv 0$, corresponding to $\rho_a = 0$. In other words, the function $\theta(x)$ which for $x = a$ has the value θ_a satisfying the first equation (40) does not in general take the value θ_b at the point $x = b$.

As a preliminary to a rigorous discussion it is useful to prove the following theorem:

COMPARISON THEOREM FOR THE FUNCTION $\theta(x)$. *Given two equations of the form* (43) *containing the same function* $p(x)$ *but different functions* $P(x)$

(44)
$$\begin{cases} \dfrac{d\theta_1}{dx} = \dfrac{1}{p(x)} \cos^2 \theta_1 + P_1(x) \sin^2 \theta_1 \\ \dfrac{d\theta_2}{dx} = \dfrac{1}{p(x)} \cos^2 \theta_2 + P_2(x) \sin^2 \theta_2 \end{cases}$$

if for all x in a certain interval (a, b), $P_1(x) < P_2(x)$, and if $\theta_1(x)$ and $\theta_2(x)$ are integrals of the two equations respectively assuming the same value θ_a for $x = a$, then $\theta_1(x) < \theta_2(x)$ for $a < x \leqq b$.*

The proof of this theorem may be based on topological considerations similar to those repeatedly used in Chapter II in investigation of the characteristics of certain first-order equations and their behaviour when crossing certain curves. Exactly what we shall prove here is that the curve $\theta = \theta_2(x)$ in the (x, θ) plane can cross exactly once (*and hence only in the point $x = a$, $\theta = \theta_a$*) the curve $\theta = \theta_1(x)$, by showing that if (x_0, θ_0) is any point of intersection of the two curves the difference $\theta_2 - \theta_1$ is necessarily *positive* to the right of x_0 and *negative* to the left of the same point.

This fact is very obvious if θ_0 is not an integral multiple of π, since then, subtracting term by term one equation in (44) from the other, we immediately obtain

$$\left[\frac{d(\theta_2 - \theta_1)}{dx} \right]_{x=x_0} = [P_2(x_0) - P_1(x_0)] \sin^2 \theta_0 > 0$$

If however $\theta_0 = n\pi$ where n is integral we may assume for convenience that $n = 0$, i.e. $\theta_0 = 0$, since equations (44) are unaltered by changing θ_1 and θ_2 into $\theta_1 - n\pi$ and $\theta_2 - n\pi$ respectively. If we now associate with the functions $\theta_1(x)$ and $\theta_2(x)$ defined by (44) and the initial conditions $\theta_1(x_0) = \theta_2(x_0) = 0$ the corresponding functions $\rho_1(x)$ and $\rho_2(x)$ defined by (37)

---

* This implies that the two integrals exist throughout the interval (a, b), which is certainly the case if, for example, the coefficients in (44) are continuous throughout this interval.

§ 22]   *An important change of variable*

where we write $a = x_0$ and for ρ_a we substitute in both cases *a like value* ρ_0, the functions

$$y_1(x) = \rho_1(x)\sin\theta_1(x), \qquad y_2(x) = \rho_2(x)\sin\theta_2(x)$$

are analogous to those functions $\bar{y}_1(x)$ and $\bar{y}_2(x)$ to which the *theorem of numerical comparison* of page 103 refers—for they satisfy two equations of type (16) containing the same function $p(x)$ and coefficients $P_1(x)$ and $P_2(x)$ such that $P_1(x) < P_2(x)$; also, at the point x_0,

$$y_1(x_0) = y_2(x_0) = 0, \qquad y_1'(x_0) = y_2'(x_0) = \frac{\rho_0}{p(x_0)}$$

since the first equation of (34) and the second of (35) yield

$$y_1'(x) = \frac{1}{p(x)}z_1(x) = \frac{\rho_1(x)}{p(x)}\cos\theta_1(x)$$

$$y_2'(x) = \frac{1}{p(x)}z_2(x) = \frac{\rho_2(x)}{p(x)}\cos\theta_2(x)$$

Hence the inequality (27) of page 103 may be applied to the functions $y_1(x)$ and $y_2(x)$, i.e. we may assert that *to the right of the point* x_0

$$y_2(x)y_1'(x) - y_1(x)y_2'(x) > 0$$

On multiplication by the positive quantity $p(x)$, it follows that

$$y_2(x)z_1(x) - y_1(x)z_2(x) > 0$$

i.e.

$$\rho_1(x)\rho_2(x)[\sin\theta_2(x)\cos\theta_1(x) - \sin\theta_1(x)\cos\theta_2(x)]$$
$$= \rho_1(x)\rho_2(x)\sin[\theta_2(x) - \theta_1(x)] > 0$$

On the contrary, *to the left of the point* x_0

$$\rho_1(x)\rho_2(x)\sin[\theta_2(x) - \theta_1(x)] < 0$$

But $\theta_1(x)$ and $\theta_2(x)$ vanish for $x = x_0$; the preceding inequalities therefore imply

$$\theta_2(x) - \theta_1(x) > 0 \quad\text{for } x > x_0, \qquad \theta_2(x) - \theta_1(x) < 0 \quad\text{for } x < x_0$$

Thus even in the more difficult case in which θ_0 is an integral multiple of π the curves $\theta = \theta_1(x)$ and $\theta = \theta_2(x)$ cannot intersect in any way other than that described earlier—and the result is established.

23. The oscillation theorem

The theorem proved in the preceding section is particularly useful when the coefficient $P(x)$ in the differential equation contains *a parameter* λ, i.e. when the equation takes the form

$$\frac{d}{dx}\left[p(x)\frac{dy}{dx}\right]+P(x,\lambda)y = 0 \tag{45}$$

This occurs in many problems of mathematical physics and we shall later discuss an example in which P takes the particular *linear* form in λ

$$P(x,\lambda) = q(x)+\lambda r(x) \tag{46}$$

If we can determine how P varies with varying λ as, for example, is known in case (46) when $r(x)$ is of constant sign, the preceding theorem enables us to determine how the function $\theta(x)$ (now more properly denoted by $\theta(x,\lambda)$) behaves for fixed x and varying λ. These results have important consequences.

We begin the discussion of the behaviour of $\theta(x,\lambda)$ with two theorems relative to the case in which $P(x,\lambda)$ tends to $\pm\infty$ as λ tends to a certain value λ^* (finite or not) of λ.

THEOREM I. *If in a certain interval* $a \leq x \leq b$

$$\lim_{\lambda \to \lambda^*} P(x,\lambda) = -\infty \tag{47}$$

and P tends uniformly to this limit, the integral $\theta(x, \lambda)$ of equation (43) satisfying the initial conditions $\theta(a, \lambda) = \alpha$ where $0 \leq \alpha < \pi$, is such that

$$\lim_{\lambda \to \lambda^*} \theta(b,\lambda) = 0 \tag{48}$$

We begin by observing that if ε is a positive number as small as we please (but always less than $\pi/2$ and less than $\pi-\alpha$) then within the strip

$$\varepsilon \leq \theta \leq \pi-\varepsilon$$

of the (x, θ)-plane, $\sin^2\theta \geq \sin^2\varepsilon$. Supposing now that λ is contained in a neighbourhood (λ', λ'') of the point λ^* so small that, by (47),

$$P(x,\lambda) < -H^2$$

where H^2 denotes a constant whose value we shall fix shortly, we deduce from (43) that if K^2 is the greatest (certainly finite) value of $1/p(x)$ in the interval (a, b)

$$\frac{d\theta}{dx} < K^2 - H^2 \sin^2 \theta \leq K^2 - H^2 \sin^2 \varepsilon$$

But this implies (see page 60) that the integral curves of equation (43) can cross only in a particular direction the integral curves of the equation

(49) $$\frac{d\theta}{dx} = K^2 - H^2 \sin^2 \varepsilon$$

(which are the lines of gradient $K^2 - H^2 \sin^2 \varepsilon$ in the (x, θ)-plane), the permissible direction being that for which the difference between the ordinate of the curve and that of the line changes from *positive* values to *negative*

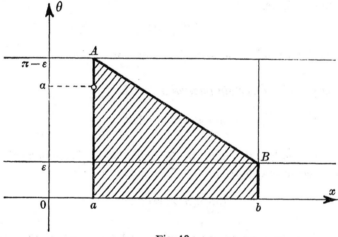

Fig. 18

values for increasing values of x. Consequently an integral curve of equation (43) which "starts off" on the lower side of one of these prescribed lines can never cross it, for increasing x, so long as it remains within the strip

$$\varepsilon \leq \theta \leq \pi - \varepsilon$$

We now fix the value of H^2 so that the specified gradient $K^2 - H^2 \sin^2 \varepsilon$ coincides with that of the line joining the two points $A(a, \pi-\varepsilon)$ and $B(b, \varepsilon)$ in figure 18, i.e. so that

$$\frac{\varepsilon - (\pi - \varepsilon)}{b - a} = K^2 - H^2 \sin^2 \varepsilon$$

which gives

$$H^2 = \frac{1}{\sin^2 \varepsilon}\left[K^2 + \frac{\pi - 2\varepsilon}{b-a}\right]$$

Therefore the integral curve referred to in the enunciation, which for $x = a$ certainly "starts off" below the point A (for since $\varepsilon < \pi - \alpha$, then $\alpha < \pi - \varepsilon$), can never cross the line-segment AB as x increases from a to b. But, on the other hand, such an integral curve can never pass into the half-plane $\theta < 0$; for, by (43), at points on the segment (a, b) of the x-axis

$$\frac{d\theta}{dx} = \frac{1}{p(x)} > 0$$

Hence this particular integral curve must remain within the shaded area in figure 18 and its ordinate for $x = b$ can be neither negative nor greater than that of the point B, i.e.

$$0 \leq \theta(b, \lambda) \leq \varepsilon$$

and as ε is arbitrary this implies the limit relation (48).

THEOREM II. *If in a certain interval $a \leq x \leq b$*

(47')
$$\lim_{\lambda \to \lambda^*} P(x, \lambda) = +\infty$$

and P tends uniformly to this limit, the integral $\theta(x, \lambda)$ of equation (43) satisfying the initial conditions $\theta(a, \lambda) = \alpha$ where $0 \leq \alpha < \pi$ is such that

(48')
$$\lim_{\lambda \to \lambda^*} \theta(b, \lambda) = +\infty$$

The proof of this second theorem is similar to that of the preceding theorem in that we construct a curve γ on which the ordinate at the point $x = b$ may be as large as we please, and below which the integral curve referred to in the enunciation can never lie.

Such a curve will be an integral curve of a differential equation (integrable by elementary methods) which *minorizes** (43). We construct it as follows:

Let σ^2 be a constant such that for $a \leq x \leq b$

$$\frac{1}{p(x)} > \sigma^2$$

---

* A similar idea is introduced in considering equation (49) as *majorizing* the same differential equation.

and τ^2 another constant to be determined later such that when λ is contained in a sufficiently restricted interval (λ', λ'') about the point λ^* then, in view of (47'),

$$P(x, \lambda) > \tau^2$$

From (43) and the fact that $\cos\theta$ and $\sin\theta$ cannot vanish simultaneously we now derive

(50) $$\frac{d\theta}{dx} > \sigma^2 \cos^2\theta + \tau^2 \sin^2\theta$$

The inequality (50) suggests the *minorizing equation*

(51) $$\frac{d\vartheta}{dx} = \sigma^2 \cos^2\vartheta + \tau^2 \sin^2\vartheta$$

which may be integrated immediately on putting

$$\tan\vartheta = \frac{\sigma}{\tau} z(x)$$

This leads to the equation in z

$$\frac{dz}{dx} = \sigma\tau(1+z^2)$$

which implies

$$z = \tan\sigma\tau(x-C)$$

where C is an arbitrary constant. The general integral of (51) is therefore

(52) $$\vartheta = \tan^{-1}\left[\frac{\sigma}{\tau}\tan\sigma\tau(x-C)\right]$$

Now take γ to be the arc in the interval (a, b) of the integral curve of (51) which issues from the point $x = a$, $\theta = 0$; this curve is defined for $a \leq x \leq b$ by (52), where C takes the value a and the inverse tangent takes values initially in the first quadrant (for $a \leq x < a+\pi/2\sigma\tau$) and subsequent values in the second, third, fourth, ..., quadrants, as follows by continuity considerations as x passes through the values $a+\pi/2\sigma\tau$, $a+\pi/\sigma\tau$, $a+3\pi/2\sigma\tau$, ... This curve rises continuously (as (51) shows that $d\vartheta/dx$ is always positive) and its ordinate assumes the values π, 2π, 3π, ..., when x assumes the values $a+\pi/\sigma\tau$, $a+2\pi/\sigma\tau$, ... ; in order to obtain the ordinate $n\pi$ (where n is an arbitrary integer) for $x = b$, let τ, which so far is undefined, be chosen (as is evidently possible) so that $a+n\pi/\sigma\tau = b$, i.e. let τ take the value

$$\tau = \frac{n\pi}{(b-a)\sigma}$$

Figure 19 illustrates the case in which $n = 4$.

But the integral curve referred to in the enunciation of the theorem which does *not* "start off" (for $x = a$) *below* the curve γ (since $\alpha \geq 0$) can

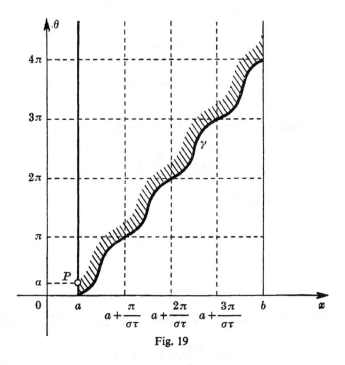

Fig. 19

never cross this curve in the interval $a<x<b$, since at a point of intersection the difference $\theta(x, \lambda) - \vartheta$ must change from *negative to positive* values (for increasing x), as from (50) and (51),

$$\frac{d(\theta - \vartheta)}{dx} > 0$$

and this is impossible; hence, within the interval (a, b), $\theta(x, \lambda) > \vartheta$ which implies

$$\theta(b, \lambda) \geq n\pi$$

In view of the arbitrariness of n the limit relation (48′) now follows.

From the preceding results, and in fact from Theorem II alone, we can immediately deduce the following theorem:

THE OSCILLATION THEOREM. *If the coefficient $P(x, \lambda)$ which appears in the differential equation (45) tends uniformly to $+\infty$ as λ tends to a certain value λ^*, finite or infinite, while x is contained in a certain interval (a, b), then, given an arbitrary positive integer n, we can always determine a neighbourhood of the point λ^* such that for any λ contained in this neighbourhood every solution of (45) vanishes at least n times within the interval (a, b).*

Under the stated conditions the function $\theta(x, \lambda)$ corresponding to any solution of (45), i.e. the solution of (43) which assumes for $x = a$ any value α in the interval $0 \leq \alpha < \pi$, attains a value as great as we please for $x = b$ (if λ is sufficiently close to λ^*) and therefore a value greater than $n\pi$. But this implies that $\theta(x, \lambda)$ which is a continuous function of x assumes at least once each of the n values

$$\pi, 2\pi, 3\pi, \ldots, n\pi$$

as x increases from a to b; hence, as $y = \rho \sin \theta$ and therefore each time that θ coincides with an integral multiple of π the function y vanishes, every solution of (45) vanishes at least n times in the interval (a, b).

24. Eigenvalues and eigenfunctions

Had we wished to prove the oscillation theorem alone the rather lengthy arguments of the last sections would be unnecessary as this result can be established by less laborious methods, for example, by comparing equation (45) with a suitable equation with constant coefficients and making use of the Sturm comparison theorem. The method adopted here has been chosen as with the three theorems of § 22 and § 23 at our disposal we are now in a position to round off neatly the discussion begun on page 110 by proving the following fundamental theorem:

THE EXISTENCE THEOREM FOR EIGENVALUES. *Given the differential equation*

(45) $$\frac{d}{dx}\left[p(x)\frac{dy}{dx}\right] + P(x, \lambda)y = 0$$

together with the boundary conditions

(6) $$hy(a) + h'y'(a) = 0, \quad ky(b) + k'y'(b) = 0$$

If $P(x, \lambda)$ is a continuous increasing function of λ in a certain interval (Λ, Λ^*) such that P tends uniformly to $-\infty$ and $+\infty$ as λ tends to Λ and Λ^* respectively, i.e.

(53) $$\lim_{\lambda \to \Lambda} P(x, \lambda) = -\infty, \quad \lim_{\lambda \to \Lambda^*} P(x, \lambda) = +\infty$$

uniformly, then there exists an infinite (increasing) sequence of values of λ,

$$\lambda_0, \lambda_1, \lambda_2, \ldots$$

contained in the interval (Λ, Λ^*), which are called EIGENVALUES of the equation (considered together with the associated boundary conditions), such that if and only if $\lambda = \lambda_n$ $(n = 0, 1, 2, 3, \ldots)$ the equation possesses solutions not identically zero satisfying the given boundary conditions. Such a solution, which is determined as far as an arbitrary multiplicative constant, vanishes exactly n times in the interval $a < x < b$ and is called an EIGENFUNCTION, or more exactly an $(n+1)^{\text{th}}$ EIGENFUNCTION.

We have seen in § 22 that the solutions of (45) corresponding to the solution $\theta(x, \lambda)$ of (43) satisfying the initial condition $\theta(a, \lambda) = \alpha$ automatically satisfy the first of the boundary conditions (6), while the second of these conditions is satisfied if and only if the second equation in (40) is satisfied, i.e. if and only if

(54) $$\theta(b, \lambda) = \beta + n\pi$$

where n is an integer and $0 < \beta \leq \pi$.

But the three theorems of § 22 and § 23 and the hypotheses of the theorem stated above imply that $\theta(b, \lambda)$ is a continuous increasing function of λ which varies from 0 to $+\infty$ as λ traverses the interval (Λ, Λ^*); therefore, for any integer $n \geq 0$, equation (54) possesses one and only one solution λ_n contained in the interval (Λ, Λ^*), and since the corresponding functions $y(x) = \rho(x) \sin \theta(x)$ vanish exactly n times within the interval (a, b) as there are exactly n integral multiples of π contained in the interval $(\alpha, \beta + n\pi)^*$, the theorem is fully established.

It is almost trivial to point out that were P a *decreasing* function of λ and were the two limits in (53) interchanged, it would be sufficient to change λ into $-\lambda$ to satisfy the conditions of the theorem; in this case there arises the sole point of difference that the sequence $\lambda_0, \lambda_1, \lambda_2, \ldots$ of eigenvalues is *decreasing*, not increasing.

---

* Note that as x increases from a to b, the function $\theta(x, \lambda)$ can assume exactly once any value which is an integral multiple of π, since for any such value of x, by (43),

$$\frac{d\theta}{dx} = \frac{1}{p(x)} > 0$$

It is easily seen that the conditions of validity of the preceding theorem are satisfied when the coefficient $P(x, \lambda)$ has the form (46) *with $r(x)$ always positive*, and $\Lambda = -\infty$, $\Lambda^* = +\infty$. If, on the other hand, $r(x) > 0$, λ must be replaced by $-\lambda$ as indicated above.

The eigenvalues and the eigenfunctions possess numerous important properties, some of which will be discussed later. Meantime we note the following:

(I) *The sequence $\lambda_0, \lambda_1, \lambda_2, \ldots$, of eigenvalues possesses the unique point of accumulation Λ^*.*

If Λ_1 is a point of accumulation, then in any neighbourhood however small of this point there exist at least two distinct eigenvalues $\lambda_{n'}$ and $\lambda_{n''}$, i.e. two values of λ such that

$$\theta(b, \lambda_{n'}) = \beta + n'\pi, \qquad \theta(b, \lambda_{n''}) = \beta + n''\pi$$

thus

$$|\theta(b, \lambda_{n'}) - \theta(b, \lambda_{n''})| = |n' - n''|\pi \geq \pi$$

which implies that the function $\theta(b, \lambda)$ is not continuous at $\lambda = \Lambda_1$. But the function $\theta(b, \lambda)$ is continuous in the interior of the interval (Λ, Λ^*); therefore either $\Lambda_1 = \Lambda$ or $\Lambda_1 = \Lambda^*$, and since Λ may be excluded as the sequence is increasing, it follows that $\Lambda_1 = \Lambda^*$.

In particular, in the case in which the coefficient $P(x, \lambda)$ is of the form (46) where $r(x) > 0$, the sole point of accumulation of the eigenvalues is the point $+\infty$ and therefore there can be only a finite number of *negative eigenvalues*.

(II) *The eigenvalues are all simple*, i.e. to each of them there corresponds exactly one eigenfunction, apart from the arbitrary multiplicative constant; for when the solution $\theta = \theta(x, \lambda_n)$ of (43) satisfying the conditions $\theta(a, \lambda_n) = \alpha$, $\theta(b, \lambda_n) = \beta + n\pi$ has been determined, the only remaining arbitrary term is the multiplicative constant ρ_a which appears in (37).*

25. A physical interpretation

The problem of boundary values for the differential equation

$$\frac{d^2y}{dx^2} + Q(x)y = 0$$

---

* This result is not trivial as with different boundary conditions from those used here, for example with the *periodicity conditions* mentioned on p. 90, the theorem may no longer be true; for corresponding to some eigenvalue λ_n the corresponding eigenfunction may depend on two arbitrary constants.

which, as we already have seen (§ 18), is no less general than equation (16), allows a physical interpretation of considerable importance, *the vibrating string*. The problem arises in the study of the vibrations of a taut string whose cross-section and whose density may vary from point to point, i.e. a string which is *not necessarily homogeneous*.

As is well known,* if we denote by $z(x, t)$ the ordinate z at time t of the centroid of the cross-section of the string corresponding to the abscissa x, under the hypothesis that at rest this centroid lies on the x-axis, we find that the function $z(x, t)$ must satisfy the partial differential equation of the second order

$$(55) \qquad \frac{\partial^2 z}{\partial x^2} - \alpha^2(x)\frac{\partial^2 z}{\partial t^2} = 0$$

where the coefficient $\alpha^2(x)$ may be expressed in terms of the area $\sigma(x)$ of the cross-section corresponding to the abscissa x, the local density $\delta(x)$ of the string and the tension τ, by the simple formula

$$(56) \qquad \alpha^2(x) = \frac{\sigma(x)\delta(x)}{\tau}$$

The equation (55) may be treated by *separation of the variables*, i.e. by seeking a solution of the form

$$(57) \qquad z(x, t) = y(x)T(t)$$

Substitution of (57) into (55) leads to the equation

$$\frac{1}{\alpha^2(x)}\frac{y''(x)}{y(x)} = \frac{T''(t)}{T(t)}$$

which can be satisfied only when both sides assume a *common constant value* which we denote by $-\lambda$. In other words, the equation (55) may be broken down into the two ordinary equations

$$(58') \qquad T''(t) + \lambda T(t) = 0$$

$$(58'') \qquad y''(x) + \lambda\alpha^2(x)y(x) = 0$$

where λ is a constant so far undetermined.

For boundary conditions it is usual to assume that the string is fixed at

---

* See, for example, F. TRICOMI (79), Part II, Chapter VIII.

its ends*—hence if l denotes its length and if its left extremity is at the origin, the boundary conditions are

$$z(0, t) = z(l, t) = 0$$

The supplementary conditions associated with (58″) are therefore

(59) $$y(0) = y(l) = 0$$

The integration of equation (58′) with constant coefficients presents no difficulties; hence the problem of the vibrating string (in which the displacement $z(x, t)$ has the form (57)) has been reduced in effect to the discussion of the Sturm-Liouville system (58″)-(59) in which, in the notation of the last section,

$$p(x) \equiv 1, \quad q(x) \equiv 0, \quad r(x) \equiv \alpha^2(x), \quad a = 0, \quad b = l$$

It is easily seen that when $q(x) \equiv 0$ and $r(x) > 0$ *there can be no negative or zero eigenvalues*, since if $\lambda \leq 0$ then $P(x) \equiv q(x) + \lambda r(x) \leq 0$, which implies (see page 98) that there exists no integral of the equation vanishing two or more times. There therefore exists an unbounded sequence of positive eigenvalues

$$\lambda_0 = \omega_0^2, \quad \lambda_1 = \omega_1^2, \quad \lambda_2 = \omega_2^2, \ldots$$

with one limit point $+\infty$, and corresponding to each of these eigenvalues the string vibrates with a frequency v_n ($n = 0, 1, 2, 3, \ldots$) determined by (58′)—for as we have already seen (page 95), the general integral of this equation when $\lambda = \omega_n^2$ is

$$T(t) = A \sin \omega_n(t - \gamma)$$

where A and γ are arbitrary constants; thus

(60) $$v_n = \frac{\omega_n}{2\pi} = \frac{1}{2\pi}\sqrt{\lambda_n} \qquad (n = 0, 1, 2, \ldots)$$

The eigenvalues therefore determine the frequencies of the harmonic modes of vibration of the string.

In particular, if the string is *homogeneous*, i.e. if $\sigma(x)\delta(x) = $ constant and therefore $\alpha(x) = \alpha = $ constant, (58″) is an equation with constant coefficients whose general integral may be written as

$$y(x) = A^* \sin \sqrt{\lambda} \alpha(x - \gamma^*)$$

---

* Other conditions on the string would imply that the supplementary conditions (59) which follow must be replaced by more general conditions of the form (6).

The conditions (59) imply

$$\gamma^* = 0, \qquad \sqrt{\lambda}\,\alpha l = (n+1)\pi$$

where n denotes a non-negative integer; thus

$$\lambda_n = \left(\frac{(n+1)\pi}{\alpha l}\right)^2, \qquad \nu_n = \frac{n+1}{2l}\sqrt{\frac{\tau}{\sigma\delta}} \qquad (n = 0, 1, 2, \ldots)$$

Thus for this problem there is a *fundamental frequency*

$$\nu_0 = \frac{1}{2l}\sqrt{\frac{\tau}{\sigma\delta}}$$

the other frequencies being multiples of this, i.e. $\nu_1 = 2\nu_0$, $\nu_2 = 3\nu_0, \ldots$. There is therefore a *fundamental note* of frequency ν_0 and *harmonics* of frequencies $2\nu_0$, $3\nu_0, \ldots$, whose various combinations give rise to the varied sounds which the vibrating string can produce.

A similar result is found in the case of the *non-homogeneous* string but with the difference that, in general, the frequencies of the harmonics are no longer integral multiples of the fundamental frequency ν_0*.

One interesting consequence of the preceding analysis is the illustration of the rather surprising fact that a sequence of discrete eigenvalues λ_0, λ_1, $\lambda_2, \ldots$, may occur in a problem such as that of the solution of a Sturm-Liouville system in which all the terms are continuous.

To illustrate this—and for greater simplicity we consider a *homogeneous string*—we draw a diagram of the displacements

$$z_n = C \sin(n+1)\frac{\pi x}{l} \cdot \sin 2(n+1)\pi\nu_0 t$$

where C is an arbitrary constant; as shown above, z_n is the displacement when the string vibrates in harmonic mode with frequency $\nu_n = (n+1)\nu_0$. We obtain from this, by putting $n = 0$, $n = 1$, $n = 2, \ldots$, figures of the types shown in figure 20, cases *a*, *b*, *c*, in each of which the various sinusoidal curves correspond to a certain fixed value of C and to equally spaced intervals for t (in cases *a* and *b*, for $\nu_0 t = \pm 1/16$, $\pm 1/8$, $\pm 3/16$, $\pm 1/4$).

When the vibrations are very rapid these figures may be regarded as visual images of the string as it appears to the eye; they show clearly that

---

* For approximations to the eigenvalues in cases other than the few special ones such as that of the homogeneous string in which an exact determination is possible, see L. COLLATZ (11) and (12); K. HOHENEMSER: 'Die Methoden zur angenährten Lösung von Eigenwertproblemen,' *Ergebnisse der Math.* I, n. 4 (Berlin, Springer, 1932); G. KRALL and R. EINAUDI: *Meccanica tecnica delle vibrazioni* (2 vols., Bologna, Zanichelli, 1940), and others.

either the cord vibrates 'all in one piece' (figure 20a where $n = 0$), or subdivides into 'two shafts' separated by a fixed point called a *node* (figure 20b where $n = 1$), or into 'three shafts' (figure 20c where $n = 2$), and so on.

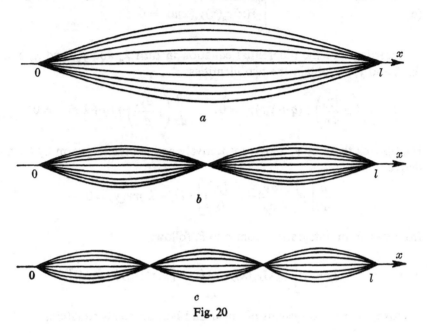

Fig. 20

As these various cases are essentially distinct, and one may not be continuously transformed into any other, it is not surprising that with these cases is associated a sequence of discrete quantities, the frequencies $v_0, v_1, v_2, \ldots$, or the eigenvalues $\lambda_0, \lambda_1, \lambda_2, \ldots$.

26. Some properties of eigenvalues and eigenfunctions

In this section we suppose, as happens in the most important concrete problems, that the coefficient $P(x, \lambda)$ in (45) depends linearly on λ, i.e. has the form

(46) $$P(x, \lambda) = q(x) + \lambda r(x)$$

and in addition that the function $r(x)$ is always *positive*. (Otherwise replace λ by $-\lambda$).

Under these conditions we shall prove that *any two eigenfunctions* $y_\mu(x)$

and $y_\nu(x)$ corresponding to the two eigenvalues λ_μ and λ_ν satisfy the fundamental ORTHOGONALITY RELATION, viz.

(61) $$\int_a^b r(x)y_\mu(x)y_\nu(x)dx = 0$$

Inserting into the differential equation in turn λ_μ and y_μ, and λ_ν and y_ν for λ and y, we obtain the two identities

$$\frac{d}{dx}\left(p\frac{dy_\mu}{dx}\right)+(q+\lambda_\mu r)y_\mu = 0, \qquad \frac{d}{dx}\left(p\frac{dy_\nu}{dx}\right)+(q+\lambda_\nu r)y_\nu = 0$$

Multiplying these by y_ν and y_μ respectively and subtracting term by term we derive

$$\frac{d}{dx}\left[p\left(y_\nu\frac{dy_\mu}{dx}-y_\mu\frac{dy_\nu}{dx}\right)\right]+(\lambda_\mu-\lambda_\nu)ry_\mu y_\nu = 0$$

from which, on integration from a to b, follows

(62) $$(\lambda_\mu-\lambda_\nu)\int_a^b r(x)y_\mu(x)y_\nu(x)dx = [p(x)(y_\mu y_\nu'-y_\nu y_\mu')]_a^b$$

Also, as the two systems of linear and homogeneous equations

$$\begin{cases} hy_\mu(a)+h'y_\mu'(a) = 0 \\ hy_\nu(a)+h'y_\nu'(a) = 0 \end{cases} \qquad \begin{cases} ky_\mu(b)+k'y_\mu'(b) = 0 \\ ky_\nu(b)+k'y_\nu'(b) = 0 \end{cases}$$

are satisfied by values of h, h' and of k, k' not simultaneously zero, the determinants of their coefficients must be zero. But these determinants are the values of

$$y_\mu(x)y_\nu'(x)-y_\nu(x)y_\mu'(x)$$

for $x = a$ and $x = b$; therefore these values must both be zero, and as $\lambda_\mu-\lambda_\nu \neq 0$, (61) follows immediately from (62).

The relation (61) shows that there exists an intimate connection between the problems dealt with in this chapter and another fundamental theory of analysis, that of *orthogonal systems of functions*. A system of functions (for example, that composed of the above eigenfunctions $y_0(x)$, $y_1(x)$, $y_2(x)$, ...) which satisfy (61) is said to be *orthogonal in the interval* (a, b), or more precisely, *orthogonal in the interval* (a, b) *with respect to the weight-function* $r(x)$. For example, the functions

$$\sin x, \quad \sin 2x, \quad \sin 3x, \quad \ldots$$

are orthogonal (with respect to the weight-function 1) in the interval $(0, \pi)$ since, if $\mu \neq \nu$,

$$\int_0^\pi \sin \mu x \sin \nu x \, dx = 0*$$

Without pursuing this subject** further, we merely observe that the eigenfunctions may be *normalized*, i.e. on multiplying by a suitable *normalizing factor* we can arrange that the integral (61) takes the value 1 when it is non-zero for $\mu = \nu$, the normalizing factor being given by

$$N_\mu = \left(\int_a^b r(x) \, y_\mu^2(x) \, dx \right)^{-\frac{1}{2}}$$

(Note that the integral under the root sign is obviously positive.)

The normalized eigenfunctions are uniquely determined apart from sign.

We now investigate by means of the comparison theorem of § 22 for the function $\theta(x)$ some important results on the *variation of the eigenfunctions for varying $q(x)$ and $r(x)$* (as they appear in (46)) while the function $p(x)$ remains unaltered. We first prove

THEOREM I. *If the function $q(x)$ increases while the functions $p(x)$ and $r(x)$ remain unaltered, the eigenvalues diminish.*

If (using the notation of the preceding section) $q^*(x)$ denotes the increased function $q(x)$, λ_n^* ($n = 0, 1, 2, \ldots$) and $\theta^*(x, \lambda)$ the eigenvalues and the function $\theta(x, \lambda)$ relative to it, then, by (54),

$$\theta(b, \lambda_n) = \theta^*(b, \lambda_n^*) = \beta + n\pi$$

On the other hand, since the relation $q^*(x) > q(x)$ implies that for the same value of λ

$$P^*(x, \lambda) = q^*(x) + \lambda r(x) > q(x) + \lambda r(x) = P(x, \lambda)$$

the comparison theorem of § 20 asserts that for the same values of x and λ

$$\theta^*(x, \lambda) > \theta(x, \lambda)$$

---

* This relation can be immediately obtained as a particular case of (61), since the functions $\sin x, \sin 2x, \ldots,$ are (as seen in the preceding section) the eigenvalues of the Sturm-Liouville system
$$y'' + \lambda y = 0, \qquad y(0) = y(\pi) = 0$$

** A book which discusses this subject thoroughly is E. C. Titchmarsh (53). In F. TRICOMI (81) the two cases of trigonometrical functions and orthogonal polynomials are dealt with.

and in particular for $x = b$, $\lambda = \lambda_n$, that

$$\theta^*(b, \lambda_n) > \theta(b, \lambda_n)$$

Hence

$$\theta^*(b, \lambda_n) > \theta^*(b, \lambda_n^*)$$

and since $\theta^*(b, \lambda)$ is an increasing function of λ this implies

$$\lambda_n^* < \lambda_n$$

as we wished to prove.

We note that the essential condition for the validity of the preceding proof is

$$P^*(x, \lambda) > P(x, \lambda)$$

Therefore the result we have derived still remains valid if $q(x)$ remains fixed while $r(x)$ increases and in addition λ_n is positive, or else while $r(x)$ decreases (but nevertheless always remains positive) and λ_n is negative. We may thus assert

THEOREM II. *If the function $r(x)$ increases while the functions $p(x)$ and $q(x)$ remain unaltered, the positive eigenvalues diminish and the negative eigenvalues increase.*

From the preceding theorems we can deduce important limits for the eigenvalues of (16) by comparing this equation with suitable equations with constant coefficients. Before we do this, however, we consider first the explicit determination of the eigenvalues of the elementary *equation of the harmonic oscillator*

(63) $$y'' + \Lambda y = 0$$

with boundary conditions taking the general form (6) rather than the special form $y(a) = y(b) = 0$ used in the previous section.

The general integral of the equation (63) is

$$y = A \sin \sqrt{\Lambda}(x - \gamma)$$

which gives

$$y' = \sqrt{\Lambda} A \cos \sqrt{\Lambda}(x - \gamma)$$

Using the equations (41), viz.

$$\alpha = \tan^{-1}_{(0,\,\pi)}\left(-\frac{h'}{h}\right), \quad \beta = \tan^{-1}_{(0,\,\pi)}\left(-\frac{k'}{k}\right)$$

the conditions (6) may be written in the form

$$\sin\sqrt{\Lambda}(a-\gamma)\cos\alpha - \sqrt{\Lambda}\cos\sqrt{\Lambda}(a-\gamma)\sin\alpha = 0$$
$$\sin\sqrt{\Lambda}(b-\gamma)\cos\beta - \sqrt{\Lambda}\cos\sqrt{\Lambda}(b-\gamma)\sin\beta = 0$$

i.e.

$$\tan\sqrt{\Lambda}(a-\gamma) = \sqrt{\Lambda}\tan\alpha, \qquad \tan\sqrt{\Lambda}(b-\gamma) = \sqrt{\Lambda}\tan\beta$$

To solve these two transcendental equations it is convenient to introduce the angles α' and β' (depending on α and β respectively and also on Λ) such that

(64) $\quad \tan\alpha' = \sqrt{\Lambda}\tan\alpha \quad (0 \leq \alpha' < \pi), \qquad \tan\beta' = \sqrt{\Lambda}\tan\beta \quad (0 < \beta' \leq \pi)$

The preceding equations then take the form

$$\tan\sqrt{\Lambda}(a-\gamma) = \tan\alpha', \qquad \tan\sqrt{\Lambda}(b-\gamma) = \tan\beta'$$

from which follow immediately

$$\sqrt{\Lambda}(a-\gamma) \equiv \alpha', \qquad \sqrt{\Lambda}(b-\gamma) \equiv \beta' \qquad (\bmod \pi)^*$$

In particular, in the case of the $(n+1)^{\text{th}}$ eigenvalue Λ_n we can write

$$\sqrt{\Lambda_n}(a-\gamma) = \alpha', \qquad \sqrt{\Lambda_n}(b-\gamma) = \beta' + n\pi$$

since this assures that $\sqrt{\Lambda_n}(x-\gamma)$ will assume exactly n times values which are integral multiples of π as x increases from a to b (extremes excluded).

Now subtracting one of these last equations from the other we derive

(65) $\qquad \sqrt{\Lambda_n} = \dfrac{\beta' - \alpha' + n\pi}{b-a}$

which, since $-\pi < \beta' - \alpha' \leq \pi$, implies

(66) $\qquad \left(\dfrac{(n-1)\pi}{b-a}\right)^2 < \Lambda_n \leq \left(\dfrac{(n+1)\pi}{b-a}\right)^2$

Now suppose, for simplicity, that the given equation is of the form

(67) $\qquad \dfrac{d^2 y}{dx^2} + [q(x) + \lambda r(x)]y = 0$

---

* The notation $A \equiv B \pmod{D}$ which stands for 'A is congruent to B modulo D' denotes that A and B differ by a multiple of D.

i.e. that the reduction (if necessary) to the form in which $p(x) = 1$ has already been carried out. We denote by q_1, q_2 and r_1, r_2 respectively the *minimum* and *maximum* values of the two functions $q(x)$ and $r(x)$ in the fundamental interval (a, b), so that

(68) $$q_1 \leq q(x) \leq q_2, \qquad 0 < r_1 \leq r(x) \leq r_2$$

From the two preceding theorems it follows that the *positive* eigenvalues diminish if we replace the functions $q(x)$ and $r(x)$ by the constants q_2 and r_2 respectively, and increase if we replace $q(x)$ and $r(x)$ by q_1 and r_1 respectively; while the *negative* eigenvalues diminish when $q(x)$ and $r(x)$ are replaced by q_2 and r_1 respectively and increase when $q(x)$ and $r(x)$ are replaced by q_1 and r_2 respectively.

Now consider the *positive eigenvalues* λ_n of (67), and in addition to (67) consider the two equations *with constant coefficients*

(69) $$\frac{d^2 y}{dx^2} + (q_1 + \lambda' r_1) y = 0, \qquad \frac{d^2 y}{dx^2} + (q_2 + \lambda'' r_2) y = 0$$

together with the same boundary conditions as has (67); if the $(n+1)^{\text{th}}$ eigenvalues of these equations are λ'_n and λ''_n respectively, then

$$\lambda''_n \leq \lambda_n \leq \lambda'_n \star$$

But by (65)

$$\sqrt{(q_1 + \lambda'_n r_1)} = \frac{\beta'_1 - \alpha'_1 + n\pi}{b-a}, \qquad \sqrt{(q_2 + \lambda''_n r_2)} = \frac{\beta'_2 - \alpha'_2 + n\pi}{b-a}$$

where α'_1, β'_1 and α'_2, β'_2 denote respectively the previous angles α', β' relative to the two equations (69); hence

(70) $$\frac{1}{r_2}\left(\frac{\beta'_2 - \alpha'_2 + n\pi}{b-a}\right)^2 - \frac{q_2}{r_2} \leq \lambda_n \leq \frac{1}{r_1}\left(\frac{\beta'_1 - \alpha'_1 + n\pi}{b-a}\right)^2 - \frac{q_1}{r_1}$$

from which, in view of

(71) $$-\pi < \beta'_1 - \alpha'_1 \leq \pi, \qquad -\pi < \beta'_2 - \alpha'_2 \leq \pi$$

$\star$ In fact the preceding theorems would allow us to replace this inequality by $\lambda''_n < \lambda_n < \lambda'_n$ provided that (68) be replaced by the stronger inequalities with $<$ in place of $\leq$. To justify the relations used here, consider two arbitrary numerical increasing sequences $q_1^{(1)}, q_1^{(2)}, q_1^{(3)}, \ldots$ and $r_1^{(1)}, r_1^{(2)}, r_1^{(3)}, \ldots$ (where the elements in the second sequence are all positive) which have as limits q_1 and r_1 respectively; then $q_1^{(k)} < q(x)$, $r_1^{(k)} < r(x)$ for all indices k. Then λ'_n can be considered as the limit as $k \to \infty$ of the $(n+1)$th eigenvalue $\lambda'_n{}^{(k)}$ of the equation corresponding to the first of (69) with coefficient $P = q_1^{(k)} + \lambda' r_1^{(k)}$. Since for every value of k, $\lambda_n < \lambda'_n{}^{(k)}$, then $\lambda_n \leq \lambda'_n$. Similarly for λ''_n.

we deduce finally

(72) $$\frac{1}{r_2}\left(\frac{(n-1)\pi}{b-a}\right)^2 - \frac{q_2}{r_2} < \lambda_n \leq \frac{1}{r_1}\left(\frac{(n+1)\pi}{b-a}\right)^2 - \frac{q_1}{r_1} \qquad (\lambda_n > 0)$$

and the equality sign can hold only in the case $\alpha = \alpha'_1 = \alpha'_2 = 0$, $\beta = \beta'_1 = \beta'_2 = \pi$, i.e. in the case of the so-called *first* boundary value problem in which the supplementary conditions are $y(a) = y(b) = 0$.

So much for the positive eigenvalues. To deal with the *negative eigenvalues* it is easily seen that all that is required is to interchange r_1 and r_2.

The double inequality (72) is of importance. For example, it shows immediately that *as n increases, λ_n increases as n^2*, i.e.

(73) $$\lambda_n = O(n^2)$$

for dividing (72) by the positive number n^2 we have

$$\frac{1}{r_2}\left[\left(1-\frac{1}{n}\right)\frac{\pi}{b-a}\right]^2 - \frac{q_2}{n^2 r_2} < \frac{\lambda_n}{n^2} \leq \frac{1}{r_1}\left[\left(1+\frac{1}{n}\right)\frac{\pi}{b-a}\right]^2 - \frac{q_1}{n^2 r_1}$$

and as $n \to \infty$ both the first and the third terms in this inequality tend to finite quantities, viz. to

$$\frac{1}{r_2}\left(\frac{\pi}{b-a}\right)^2 \quad \text{and} \quad \frac{1}{r_1}\left(\frac{\pi}{b-a}\right)^2 \quad \text{respectively}$$

This proves the result.

Further, if $r_1 = r_2$, i.e. *if the function $r(x)$ reduces to the constant r_1*, we may assert immediately

$$\lim_{n\to\infty} \frac{\lambda_n}{n^2} = \frac{1}{r_1}\left(\frac{\pi}{b-a}\right)^2$$

or better, that

$$\frac{\lambda_n}{n^2} = \frac{1}{r_1}\left(\frac{\pi}{b-a}\right)^2 + O\left(\frac{1}{n}\right)$$

i.e.

(74) $$\lambda_n = \frac{1}{r_1}\left(\frac{\pi}{b-a}\right)^2 n^2 + O(n)$$

From the many corollaries that may be deduced from these asymptotic relations we note only the following:

The series $\Sigma 1/\lambda_n$ of inverses of the eigenvalues of a Sturm-Liouville system is always convergent.

Equation (73) implies $1/\lambda_n = O(n^{-2})$; hence if A is an upper bound of λ_n^{-1}/n^{-2} the series $\Sigma 1/\lambda_n$ is majorized by the convergent series $\Sigma A/n^2$.

We now discuss briefly the auxiliary angles α' and β' which appear in the explicit expression (65) for the eigenvalues of an equation with constant coefficients, and in (70).

Provided that $\alpha \neq 0$ the first equation (64) may be written in the form

$$\tan\left(\frac{\pi}{2} - \alpha'\right) = \frac{1}{\tan \alpha'} = \frac{1}{\sqrt{\Lambda} \tan \alpha} = \frac{\cot \alpha}{\sqrt{\Lambda}}$$

so that

$$\alpha' = \frac{\pi}{2} - \tan^{-1} \frac{\cot \alpha}{\sqrt{\Lambda}}$$

Now expanding the inverse tangent (which is certainly permissible if $\Lambda > \cot^2 \alpha$) we have

(75) $$\alpha' = \frac{\pi}{2} - \frac{\cot \alpha}{\sqrt{\Lambda}} + \frac{1}{3}\left(\frac{\cot \alpha}{\sqrt{\Lambda}}\right)^3 - \ldots$$

Similarly (supposing that $\beta \neq \pi$ and $\Lambda > \cot^2 \beta$)

(75') $$\beta' = \frac{\pi}{2} - \frac{\cot \beta}{\sqrt{\Lambda}} + \frac{1}{3}\left(\frac{\cot \beta}{\sqrt{\Lambda}}\right)^3 - \ldots$$

These last formulae may in the case of *differential equations with constant coefficients* be used to obtain approximations to the eigenvalues in place of the asymptotic formula (74), and in fact the formulae (75)-(75') give sharper results, i.e. the *residues* are of a lower order.

As $\Lambda_n = O(n^2)$ it follows from (75) that, provided α is not zero, the quantity α' in (65) is given by

$$\alpha' = \frac{\pi}{2} - \frac{\cot \alpha}{\sqrt{\Lambda_n}} + O(n^{-3})$$

As (74), in which we now put $r_1 = 1$, yields*

$$\frac{1}{\sqrt{\Lambda_n}} = \frac{b-a}{\pi n}[1+O(n^{-1})]^{-\frac{1}{2}} = \frac{b-a}{\pi n} + O(n^{-2})$$

this therefore gives

$$\alpha' = \frac{\pi}{2} - \frac{b-a}{\pi n}\cot \alpha + O(n^{-2})$$

---

* To justify the equality $[1 + O(n^{-1})]^{-1/2} = 1 + O(n^{-1})$ without using an expansion in series see the following § 34.

Similarly, if $\beta \neq \pi$,

$$\beta' = \frac{\pi}{2} - \frac{b-a}{\pi n}\cot\beta + O(n^{-2})$$

We require only to substitute these formulæ into (65) to deduce that

(1) If $\alpha \neq 0$, $\beta \neq \pi$,

(76) $$\sqrt{\Lambda_n} = n\frac{\pi}{b-a} + \frac{1}{n\pi}(\cot\alpha - \cot\beta) + O(n^{-2})$$

(2) If $\alpha \neq 0$, $\beta = \pi$ (which implies that also $\beta' = \pi$),

(76') $$\sqrt{\Lambda_n} = (n+\tfrac{1}{2})\frac{\pi}{b-a} + \frac{1}{n\pi}\cot\alpha + O(n^{-2})$$

(3) If $\alpha = 0$ (which implies that also $\alpha' = 0$), $\beta \neq \pi$,

(76'') $$\sqrt{\Lambda_n} = (n+\tfrac{1}{2})\frac{\pi}{b-a} - \frac{1}{n\pi}\cot\beta + O(n^{-2})$$

(4) If $\alpha = 0$, $\beta = \pi$ (which implies that also $\alpha' = 0$, $\beta' = \pi$),

(76''') $$\sqrt{\Lambda_n} = (n+1)\frac{\pi}{b-a}$$

in conformity (apart from notation) with the results on page 122.

To show that (76) represents an improvement on the formula (74), it is only necessary to square; this gives

(77) $$\Lambda_n = \left(\frac{\pi}{b-a}\right)^2 n^2 + \frac{2}{b-a}(\cot\alpha - \cot\beta) + O(n^{-1})$$

while (74) has second term $O(n)$. It is preferable, however, either to leave the formula in the form (76) or write it as

(76*) $$\sqrt{\Lambda_n} = n\frac{\pi}{b-a} + \frac{1}{n\pi}\left(\frac{k}{k'} - \frac{h}{h'}\right) + O(n^{-2})$$

since

(78) $$\cot\alpha = \frac{1}{\tan\alpha} = -\frac{h}{h'}, \quad \cot\beta = \frac{1}{\tan\beta} = -\frac{k}{k'}$$

In the following chapter we shall see how the preceding formulæ may be extended to equations whose coefficients are not constant.

By comparing a given equation with other suitable equations with non-constant coefficients whose eigenvalues, in some way, are known, it is sometimes possible to obtain such sharp limits for the eigenvalues of the given equation (by using the theorems of pages 125-6) that a fairly good approximation to its eigenvalues actually results.

27. Connection with the theory of integral equations

In the previous section we have seen how the questions dealt with in the present chapter are connected with the theory of orthogonal systems of functions. We now show how they tie up with another fundamental theory of modern analysis, viz. the theory of *integral equations*. It will appear that the integral equations concerned are *not of Volterra type* as were the integral equations which appeared in Chapter I but are *of Fredholm type*, i.e. equations involving integrals between constant limits which may be *non-homogeneous* as

(79) $$\phi(x) - \lambda \int_a^b K(x, y) \phi(y) \, dy = f(x)$$

or *homogeneous* as

(80) $$\phi(x) - \lambda \int_a^b K(x, y) \phi(y) \, dy = 0$$

where K, f, a, and b are supposed given, while ϕ and possibly λ are to be determined.*

The link referred to is to be found in consideration of the *Green's function* of a given Sturm-Liouville system. We shall consider here the usual boundary conditions (6) and the equation (16) with coefficient $P = q(x)$, viz.

(81) $$\frac{d}{dx}\left[p(x)\frac{dy}{dx}\right] + q(x) y = 0$$

i.e. we suppose at first that the λ which appears in (46) is *zero*, and further that

(I) $\lambda = 0$ is *not* an eigenvalue of the system, i.e. there exists no solution not identically zero of (81) satisfying the boundary conditions (6);

(II) we can find two linearly independent integrals $y_1(x)$ and $y_2(x)$ of (81), as, for example, can certainly be done when $q(x) \equiv 0$.

---

* These equations are called equations of the *second kind*, in contrast with equations of the *first kind* in which no term $\phi(x)$ appears outside the integral sign. (Equations of the first kind will not arise in this book.)

The impossibility of determining a solution not identically zero of (81) satisfying the boundary conditions (6) is evidently due to the demand that it be a *continuous function with continuous first derivative*, which implies (assuming that p and q can be differentiated as often as required) that its successive derivatives are also continuous. In fact, if we drop one or other of the last two conditions, for example that the first derivative be continuous, it is then possible to construct, as we shall show, a solution of (81) which is not identically zero and which satisfies the boundary conditions (6); further we may even prescribe the *magnitude of the jump* in its first derivative at the unique point in the interval (a, b) at which this derivative is not continuous.

To do this we denote by $U(x)$ and $V(x)$ two known particular integrals (not identically zero) of (81) satisfying respectively the first and second conditions (6); as the general integral of (81) is $c_1 y_1(x) + c_2 y_2(x)$ we obtain for $U(x)$ and $V(x)$ the following expressions*

(82) $$\begin{cases} U(x) = [hy_2(a) + h'y_2'(a)]y_1(x) - [hy_1(a) + h'y_1'(a)]y_2(x) \\ V(x) = [ky_2(b) + k'y_2'(b)]y_1(x) - [ky_1(b) + k'y_1'(b)]y_2(x) \end{cases}$$

Now let ξ be any internal point of the interval (a, b) and A and B be two constants not both zero yet to be specified. Consider the function $G(x, \xi)$ defined by

(83) $$G(x, \xi) = \begin{cases} AU(x) & (a \leq x < \xi) \\ BV(x) & (\xi < x \leq b) \end{cases}$$

Such a function, in general discontinuous at the point ξ, satisfies the following conditions: (a) it is not identically zero; (b) it satisfies the differential equation at every point x of the interval (a, b) other than the point ξ; (c) it satisfies both boundary conditions (6). In addition, in view of the arbitrariness of the constants A and B we may further demand (d) that the function G is continuous also at the point ξ; (e) that the derivative of the function G *jumps* a prescribed amount $1/p(\xi)$** at the point $x = \xi$, i.e.

(84) $$G_x(\xi+0, \xi) - G_x(\xi-0, \xi) = \frac{1}{p(\xi)}$$

---

* These exclude the possibility that either U or V is identically zero; for if, for example,
$$\begin{cases} hy_1(a) + h'y_1'(a) = 0 \\ hy_2(a) + h'y_2'(a) = 0 \end{cases}$$
then, since the Wronskian of y_1 and y_2 is not zero at the point $x = a$, $h = h' = 0$, contrary to hypothesis.

** Many authors define Green's function by specifying the jump in its derivative to be $-1/p(\xi)$; this implies here the change of G into $-G$.

In order that these last two conditions may be satisfied the constants A and B must be chosen so that

(85) $$\begin{cases} AU(\xi) - BV(\xi) = 0 \\ AU'(\xi) - BV'(\xi) = -\dfrac{1}{p(\xi)} \end{cases}$$

This is possible since this linear (non-homogeneous) system has one and only one solution, as the determinant of its coefficients is *not zero*—since otherwise it would be possible to determine a G_x continuous at the point $x = \xi$ which would imply $AU'(\xi) - BV'(\xi) = 0$, contrary to the hypothesis that $\lambda = 0$ is *not* an eigenvalue.

The function $G(x, \xi)$ satisfying the five conditions a, b, c, d, and e above is called Green's function for the given Sturm-Liouville system, i.e. the function, continuous but with discontinuous first derivative, defined by (83), where the constants A and B have the definite values determined by equations (85).

To show how the use of Green's function in the boundary-values problem for the equation

(86) $$\frac{d}{dx}\left[p(x)\frac{dy}{dx}\right] + [q(x) + \lambda r(x)]y = 0$$

together with the conditions (6) reduces the system to an integral equation of Fredholm type, it is convenient to reproduce the reduction already carried out on page 102 and again on page 124; if we write

$$\frac{d}{dx}\left[p(x)\frac{dy}{dx}\right] + q(x)y \equiv L(y)$$

(86) may be written as

(86') $$L(y) + \lambda r(x)y = 0$$

whence

(87) $$zL(y) - yL(z) = \frac{d}{dx}\left[p\left(z\frac{dy}{dx} - y\frac{dz}{dx}\right)\right]$$

It is important here to note that the integral from a to b of the derivative of a function $F(x)$ with a discontinuity (or perhaps more than one) at a point (points) within (a, b) is not given as in the ordinary case by $F(b) - F(a)$, but by this difference *minus the sum of the jumps of F(x) between a and b*.*

On identifying the function $z(x)$ in (87) with $G(x, \xi)$, so that $L(z) \equiv 0$, and integrating (87) from a to b, we obtain

$$\int_a^b G(x, \xi) L(y) \, dx = \left[p\left(G\frac{dy}{dx} - y\frac{\partial G}{\partial x} \right) \right]_a^b - \lim_{\varepsilon \to 0} \left[p\left(G\frac{dy}{dx} - y\frac{\partial G}{\partial x} \right) \right]_{\xi-\varepsilon}^{\xi+\varepsilon}$$

Also both functions y and G satisfy the boundary conditions (6); hence (cf. p. 124)

$$\left[p\left(G\frac{dy}{dx} - y\frac{\partial G}{\partial x} \right) \right]_a^b = 0$$

Also, in view of (84),

$$\lim_{\varepsilon \to 0} \left[p\left(G\frac{dy}{dx} - y\frac{\partial G}{\partial x} \right) \right]_{\xi-\varepsilon}^{\xi+\varepsilon}$$

$$= -\lim_{\varepsilon \to 0} \left[p(\xi+\varepsilon) y(\xi+\varepsilon) G_x(\xi+\varepsilon, \xi) - p(\xi-\varepsilon) y(\xi-\varepsilon) G_x(\xi-\varepsilon, \xi) \right]$$

$$= -p(\xi) y(\xi) \lim_{\varepsilon \to 0} \left[G_x(\xi+\varepsilon, \xi) - G_x(\xi-\varepsilon, \xi) \right] = -y(\xi)$$

whence

(88) $$\int_a^b G(x, \xi) L(y) \, dx = y(\xi)$$

which yields, in view of (86'), the integral equation of type (80),

(89) $$y(\xi) + \lambda \int_a^b G(x, \xi) r(x) y(x) \, dx = 0$$

This completes the proof that the *differential equation* (86) *with the boundary conditions* (6) *is equivalent to the homogeneous Fredholm integral equation* (89), which therefore, supposing $r(x) > 0$, possesses solutions not identically zero (and determined apart from a multiplicative constant) if and only if λ is an eigenvalue, i.e. λ coincides with one of the elements of a known unbounded sequence $\lambda_0, \lambda_1, \lambda_2, \ldots$, having as unique point of accumulation the point $+\infty$.

Conversely, it may be easily shown that every solution of (89) satisfies equation (86) and the boundary conditions (6).

* If, for example, the function F has one point of discontinuity c within the interval (a, b) then

$$\int_a^b F'(x) \, dx = \lim_{\varepsilon \to 0} \int_a^{c-\varepsilon} + \int_{c+\varepsilon}^b F'(x) \, dx = F(b) - F(a) - \lim_{\varepsilon \to 0} [F(c+\varepsilon) - F(c-\varepsilon)]$$

Cf. F. TRICOMI (79), Part II, Chapter I, § 12.

To those readers with a knowledge of the theory of Fredholm integral equations the above statements may seem in contradiction with the fact that such equations do not always possess eigenvalues; for example, the equations of Volterra type which appear in (79)–(80) when the *nucleus* (or *kernel*) $K(x, y)$ is identically zero for $y > x$, do not possess eigenvalues.

This apparent contradiction may be removed by showing that equation (89) may be reduced to an equation with *symmetrical nucleus*, i.e. such that $K(x, y) \equiv K(y, x)$, for such equations always possess eigenvalues.*

We begin by showing that the *Green's function is symmetrical in both its arguments*.

Again consider the identity (87) and put

$$y = G(x, \xi), \quad z = G(x, \eta)$$

where ξ and η are any two distinct points whatsoever in the interval (a, b). As $L(y) \equiv L(z) \equiv 0$, the identity (87) reduces to

$$\frac{d}{dx}\{p(x)[G(x, \eta)G_x(x, \xi) - G(x, \xi)G_x(x, \eta)]\} = 0$$

and on integrating from a to b (noting that there are this time two points of discontinuity, namely $x = \xi$ and $x = \eta$) we have

$$\lim_{\varepsilon \to 0}\{p(x)[G(x, \eta)G_x(x, \xi) - G(x, \xi)G_x(x, \eta)]\}_{\xi-\varepsilon}^{\xi+\varepsilon}$$
$$+ \lim_{\varepsilon \to 0}\{p(x)[G(x, \eta)G_x(x, \xi) - G(x, \xi)G_x(x, \eta)]\}_{\eta-\varepsilon}^{\eta+\varepsilon}$$
$$= \{p(x)[G(x, \eta)G_x(x, \xi) - G(x, \xi)G_x(x, \eta)]\}_a^b = 0$$

But the first limit is obviously equal to

$$\lim_{\varepsilon \to 0}[p(x)G(x, \eta)G_x(x, \xi)]_{\xi-\varepsilon}^{\xi+\varepsilon} = p(\xi)G(\xi, \eta)\lim_{\varepsilon \to 0}[G_x(x, \xi)]_{\xi-\varepsilon}^{\xi+\varepsilon} = G(\xi, \eta)$$

while the second limit is equal to $-G(\eta, \xi)$; therefore

$$G(\xi, \eta) - G(\eta, \xi) = 0$$

as we aimed to prove.

Also, as $r(x) > 0$, by multiplying equation (89) by $\sqrt{r(\xi)}$ we have

$$\sqrt{[r(\xi)]}\, y(\xi) + \lambda \int_a^b G(x, \xi)\sqrt{[r(x)]}\sqrt{[r(\xi)]}\sqrt{[r(x)]}\, y(x)\, dx = 0$$

---

* A useful and concise account of the theory of (linear) integral equations is contained in R. COURANT-D. HILBERT (59), Chapter III; also in M. PICONE (75); also in G. HAMEL: *Integralgleichungen* (2nd edition, Berlin, Springer (1949)). A deeper account is to be found in the author's treatise (82).

If now we put

(90) $$\sqrt{[r(x)]}\,y(x) = \phi(x), \qquad -G(x,\xi)\sqrt{[r(x)r(\xi)]} = K(x,\xi)$$

the integral equation obtained earlier may be replaced by the following one with *symmetrical nucleus*,

(91) $$\phi(\xi) - \lambda \int_a^b K(x,\xi)\,\phi(x)\,dx = 0$$

This transformation allows us to use, for Sturm-Liouville systems, the various methods so far devised for the approximate determination of the eigenvalues of a Fredholm equation with *symmetrical nucleus*; particularly useful among these is the *variational method* based on the study of the maxima of suitable integrals. For information on these methods we refer the reader to the works cited on page 122 and to the book (82) by the author; also to the remark at the end of the previous section (p. 132).

We note in conclusion that by using Green's function the integral of the linear *non-homogeneous* equation

(92) $$\frac{d}{dx}\left[p(x)\frac{dy}{dx}\right] + q(x)y = f(x)$$

satisfying given *homogeneous** conditions at the extremities of an interval (a, b) may be put into a very elegant form without employing the Lagrange method of variation of parameters (provided that the integration of the corresponding homogeneous equation can be carried out).

In this case we have $L(y) = f(x) =$ a known function, whence, from (88), follows

(93) $$y(\xi) = \int_a^b G(x,\xi)f(x)\,dx$$

where $G(x,\xi)$ is the Green's function of the corresponding homogeneous equation for the same given boundary conditions.

---

* If the conditions are *non-homogeneous*:

$$hy(a) + h'y'(a) = \alpha, \qquad ky(b) + k'y'(b) = \beta$$

and if $\eta(x)$ be *any* function which satisfies these conditions, write $z(x) = y(x) - \eta(x)$; this change of variable transforms (92) into a similar equation in $z(x)$, the sole difference being that $f(x)$ is now replaced by

$$f(x) - \frac{d}{dx}\left[p(x)\frac{d\eta}{dx}\right] - q(x)\eta$$

This substitution reduces the problem to one with the boundary conditions of the form (6).

As has already been pointed out the reduction to an integral equation is particularly useful when we are dealing with boundary conditions more general than those of (6). By this method we can also treat the case which arises in many applications in which the boundary conditions involve the same parameter λ as appears in the equation; for this we need to consider an equation with nucleus a function $K(x, \xi, \lambda)$. Such equations have already been much studied and we mention in particular the work of D. Greco.*

* D. GRECO: *G. di Mat. di Battaglini*, **78**, 216–37 (1948–49); 79, 86–120 (1949–50).

IV. Asymptotic Methods

28. General remarks

In recent years much use has been made of *asymptotic methods* in many branches of analysis and in particular in the study of differential equations; these are methods which describe the behaviour of certain functions when some of the independent variables on which they depend tend to infinity or to some particular values. For example, when in § 26 we showed that the $(n+1)$th eigenvalue of the differential equation

$$\frac{d^2y}{dx^2}+[q(x)+\lambda r(x)]y = 0$$

with general homogeneous boundary conditions at the extremes of the interval (a, b) and for which $r(x) = r_1 =$ constant, is such that

(1) $$\lambda_n = \frac{1}{r_1}\left(\frac{\pi}{b-a}\right)^2 n^2 + O(n), \quad n \to \infty$$

we in fact used asymptotic methods.

The importance of results of this kind is indicated, for example, by the fact that from (1), or from the less exact asymptotic formula

$$\lambda_n = O(n^2)$$

of § 26, we have been able to prove that the series $\Sigma 1/\lambda_n$ is always convergent. In addition, formulæ of type (1) often yield useful approximations, as in some cases a sufficiently good approximation to the value of the left-hand side is obtained by considering only the explicit terms on the right-hand side.

In the theory of ordinary differential equations asymptotic methods are applied principally to the following three types of problems:

(1) The determination, for example by means of formulæ of type (1), of the behaviour of all or of some of the integrals of a given differential

equation as the independent variable x tends to infinity or to some particular value.*

(2) The determination of the behaviour of certain integrals $y(x)$ of the equation, or of quantities defined by these, when x remains unaltered while some parameter contained in the equation or in the supplementary conditions tends to infinity or to some particular value.

(3) The determination of the particular integral or integrals of the equation which behave asymptotically in a prescribed manner, for example which always remain bounded when x tends to infinity.

The problems treated in the last chapter may be considered as belonging to the second group.

Among the problems of the first group the following one is of predominating importance. For greater clarity and simplicity we shall consider the case of a linear equation of the second order but the problem can obviously be generalized.

Given the differential equation

$$(2) \qquad y'' + p_1(x)y' + p_2(x)y = 0$$

where

$$(3) \qquad \lim_{x \to \infty} p_1(x) = a_0, \qquad \lim_{x \to \infty} p_2(x) = b_0$$

can we assert that its integrals for very large values of x behave similarly to those of the *limit equation*

$$(4) \qquad y'' + a_0 y' + b_0 y = 0$$

i.e. similarly to some linear combination of the two exponential functions

$$(5) \qquad e^{\alpha_1 x}, \qquad e^{\alpha_2 x}$$

where α_1 and α_2 are the roots (which for simplicity we suppose distinct) of the characteristic equation

$$(6) \qquad \alpha^2 + a_0 \alpha + b_0 = 0$$

of (4)?

---

* The second case reduces to the first on making the change of variable

$$\xi = \frac{1}{(x - x_0)}; \quad \text{then } \xi \to \infty \text{ as } x \to x_0.$$

No answer can be given to this question without further consideration.

As illustration of this remark we need only consider the particular case in which the functions $p_1(x)$ and $p_2(x)$ are *analytic functions regular at infinity*, i.e. they may be represented for sufficiently large $|x|$ by series of the form

(7) $$p_1(x) = a_0 + \frac{a_1}{x} + \frac{a_2}{x^2} + \dots, \qquad p_2(x) = b_0 + \frac{b_1}{x} + \frac{b_2}{x^2} + \dots$$

and consequently the relations (3) are satisfied.

We shall see in the next chapter that under these hypotheses we can find two linearly independent integrals of (2) representable (in a sense that will be defined) by series of the form

$$e^{\alpha_1 x} x^{\rho_1} \left(1 + \frac{c_1'}{x} + \frac{c_2'}{x^2} + \dots \right), \qquad e^{\alpha_2 x} x^{\rho_2} \left(1 + \frac{c_1''}{x} + \frac{c_2''}{x^2} + \dots \right)$$

where the two exponents ρ_1 and ρ_2 are given by the formulæ

(8) $$\rho_1 = -\frac{a_1 \alpha_1 + b_1}{a_0 + 2\alpha_1}, \qquad \rho_2 = -\frac{a_1 \alpha_2 + b_1}{a_0 + 2\alpha_2}$$

The *asymptotic behaviour* of the integrals of (2) is therefore not analogous to that of a linear combination of the exponential functions (5) but to that of a linear combination of the functions

(9) $$e^{\alpha_1 x} x^{\rho_1}, \qquad e^{\alpha_2 x} x^{\rho_2}$$

which, in view of (8), do not depend solely on a_0 and b_0, i.e. do not depend solely on the values to which $p_1(x)$ and $p_2(x)$ tend as $x \to \infty$, but also on the coefficients a_1 and b_1 appearing on the right-hand sides of (7). However, as in general an exponential factor is more powerful than a power, the above work suggests that *in many cases* valid results on the behaviour of the integrals of (2) as $x \to \infty$ may be derived from the limit equation (4), i.e. ultimately from study of the roots of the characteristic equation (6). But this requires rigorous proof and is not at all evident as seems to be popularly believed.*

---

* Several of the results so far derived in this work for, by example, Dini, have been established under hypotheses which imply in the analytic case that $a_1 = b_1 = 0$. This explains why there is no contradiction between these results and what has been said above.

29. A general method applicable to linear differential equations

In researches on asymptotic behaviour, and also for study of other properties, an extremely useful method is that which we choose to call *the method of Fubini**; this consists in 'comparing' a given differential equation by means of an integral equation of Volterra type with a suitable 'approximate' equation which can be integrated explicitly or for which some special properties of the integrals are known.

Suppose, for simplicity, that we are dealing with a linear and homogeneous equation of the second order** which we suppose written in the form

(10) $$y'' + p_1(x)y' + p_2(x)y = A(x)y'' + B(x)y' + C(x)y$$

This can evidently be done in an infinity of ways, even when the restriction is added that the 'approximate' equation

(11) $$y'' + p_1(x)y' + p_2(x)y = 0$$

must be either explicitly integrable or else such that two of its linearly independent integrals

(12) $$y_1 = F_1(x), \quad y_2 = F_2(x)$$

possess some special property, for example that they are *stable*, i.e. bounded, for $x \to \infty$.

The method consists in *formally* considering (10) as a non-homogeneous equation and applying to it Lagrange's method of variation of parameters; this leads us to put

(13) $$y = C_1(x)F_1(x) + C_2(x)F_2(x)$$

where $C_1(x)$ and $C_2(x)$ are two unknown functions such that

(14) $$C_1'(x)F_1(x) + C_2'(x)F_2(x) = 0$$

---

* It has been pointed out to me that the historical reasons for attributing this method to Fubini are rather weak, for the method used by Fubini in 1937 (*Rend. Lincei*, (6) 26, 253–59) to study the asymptotic behaviour of Bessel functions is not substantially different from the method of Liouville-Stekloff which has recently been improved by R. E. Langer. (See, for example, G. Szegö (78), p. 204.) It seems that the *general* method (applicable when the right-hand side of (10) contains terms in y' and y'') appeared for the first time in the first edition of this book and in a related publication of the author [*Rend. Seminario Mat. Torino*, **8**, 7–19 (1947–48)]. I have however continued to speak of the *method of Fubini* in memory of my dear friend and colleague G. Fubini (1879–1943).

** The method may be extended easily to linear equations of order n and even to some types of non-linear equations.

These equations imply

$$y' = C_1(x) F_1'(x) + C_2(x) F_2'(x)$$
$$y'' = C_1(x) F_1''(x) + C_2(x) F_2''(x) + C_1'(x) F_1'(x) + C_2'(x) F_2'(x)$$

Now substituting into (10) and using the fact that F_1 and F_2 are two particular integrals of (11) we obtain

$$C_1'F_1' + C_2'F_2' =$$
$$= C_1(AF_1'' + BF_1' + CF_1) + C_2(AF_2'' + BF_2' + CF_2) + A(C_1'F_1' + C_2'F_2')$$

and therefore supposing, as is obviously true, that $A \neq 1$,

$$C_1'F_1' + C_2'F_2' = \frac{AF_1'' + BF_1' + CF_1}{1-A} C_1 + \frac{AF_2'' + BF_2' + CF_2}{1-A} C_2$$

This equation together with (14) gives a system of two linear equations in the two unknowns C_1' and C_2' for which the determinant of the coefficients is *not zero* as it coincides with the *Wronskian*

$$W(x) = \begin{vmatrix} F_1(x) & F_2(x) \\ F_1'(x) & F_2'(x) \end{vmatrix}$$

of the two integrals (12). There is therefore no difficulty in solving such a system; writing for convenience

(15) $$\mathscr{G}_h(x) = \frac{A(x) F_h''(x) + B(x) F_h'(x) + C(x) F_h(x)}{[1-A(x)] W(x)} \qquad (h = 1, 2)$$

we find

(16) $$C_1' = -F_2(\mathscr{G}_1 C_1 + \mathscr{G}_2 C_2), \qquad C_2' = F_1(\mathscr{G}_1 C_1 + \mathscr{G}_2 C_2)$$

Hence if x_0 is any value whatsoever of x, and γ_1 and γ_2 two arbitrary constants representing respectively the values of $C_1(x)$ and $C_2(x)$ for $x = x_0$, the functions $C_1(x)$ and $C_2(x)$ must satisfy a system of two linear integral equations of Volterra type, namely

(17) $$\begin{cases} C_1(x) = \gamma_1 - \int_{x_0}^{x} F_2(\xi) [\mathscr{G}_1(\xi) C_1(\xi) + \mathscr{G}_2(\xi) C_2(\xi)] d\xi \\ C_2(x) = \gamma_2 + \int_{x_0}^{x} F_1(\xi) [\mathscr{G}_1(\xi) C_1(\xi) + \mathscr{G}_2(\xi) C_2(\xi)] d\xi \end{cases}$$

Multiplying these respectively by $\mathscr{G}_1(x)$ and $\mathscr{G}_2(x)$ and adding, we obtain the single integral equation

$$(18) \qquad \phi(x) - \int_{x_0}^{x} \begin{vmatrix} F_1(\xi) & F_2(\xi) \\ \mathscr{G}_1(x) & \mathscr{G}_2(x) \end{vmatrix} \phi(\xi)\, d\xi = \gamma_1 \mathscr{G}_1(x) + \gamma_2 \mathscr{G}_2(x)$$

containing the one unknown function

$$\phi(x) = \mathscr{G}_1(x)\, C_1(x) + \mathscr{G}_2(x)\, C_2(x)$$

Further, if we write

$$\phi(x) = \gamma_1 \phi_1(x) + \gamma_2 \phi_2(x)$$

the arbitrary constants disappear from the integral equation, for (18) breaks down into the two single equations

$$(19) \qquad \phi_h(x) - \int_{x_0}^{x} K(x, \xi)\, \phi_h(\xi)\, d\xi = \mathscr{G}_h(x) \qquad (h = 1, 2)$$

where

$$(20) \qquad K(x, \xi) = \begin{vmatrix} F_1(\xi) & F_2(\xi) \\ \mathscr{G}_1(x) & \mathscr{G}_2(x) \end{vmatrix}$$

Hence on solving (19) and employing (13) and (17), we find

$$y = \gamma_1 F_1(x) + \gamma_2 F_2(x) + \int_{x_0}^{x} L(x, \xi)\, [\gamma_1 \phi_1(\xi) + \gamma_2 \phi_2(\xi)]\, d\xi$$

where

$$(21) \qquad L(x, \xi) = \begin{vmatrix} F_1(\xi) & F_2(\xi) \\ F_1(x) & F_2(x) \end{vmatrix}$$

In other words,

$$(22) \qquad y(x) = \gamma_1 Y_1(x) + \gamma_2 Y_2(x)$$

where Y_1 and Y_2 are the functions (independent of the constants γ_1 and γ_2) given by the formulæ

$$(23) \qquad Y_h(x) = F_h(x) + \int_{x_0}^{x} L(x, \xi)\, \phi_h(\xi)\, d\xi \qquad (h = 1, 2)$$

The reduction of the problem to integral equations may be carried out in a rather different way which has several advantages. For example (as we shall see below), it leads to a single integral equation in place of two, in which the unknown function is the same $y(x)$ as is to be determined instead of some unknown auxiliary functions such as $\phi_1(x)$ and $\phi_2(x)$. In addition, in several cases as for example in that of Liouville-Stekloff in which $A(x) \equiv B(x) \equiv 0$ the nucleus of the new equation may well be simpler than the preceding nucleus $K(x, \xi)$.

The new reduction is obtained by adding the equations (17), having first multiplied these equations by $F_1(x)$ and $F_2(x)$ respectively in place of $\mathscr{G}_1(x)$ and $\mathscr{G}_2(x)$; this gives the equation

$$y(x) = \gamma_1 F_1(x) + \gamma_2 F_2(x) + \int_{x_0}^{x} L(x, \xi) \phi(\xi) \, d\xi$$

in which, in view of (15) and since F_1 and F_2 are particular integrals of (11), we may replace $\phi(\xi)$ by the expression

$$\phi(\xi) = \sum_{h=1}^{2} \frac{A(\xi)[-p_1(\xi) F_h'(\xi) - p_2(\xi) F_h(\xi)] + B(\xi) F_h'(\xi) + C(\xi) F_h(\xi)}{[1 - A(\xi)] W(\xi)} C_h(\xi)$$

$$= \frac{B(\xi) - A(\xi) p_1(\xi)}{[1 - A(\xi)] W(\xi)} \sum_{h=1}^{2} C_h(\xi) F_h'(\xi) + \frac{C(\xi) - A(\xi) p_2(\xi)}{[1 - A(\xi)] W(\xi)} \sum_{h=1}^{2} C_h(\xi) F_h(\xi)$$

$$= M(\xi) y(\xi) + N(\xi) y'(\xi)$$

where

(24) $\quad M(x) = \dfrac{C(x) - A(x) p_2(x)}{[1 - A(x)] W(x)}, \quad N(x) = \dfrac{B(x) - A(x) p_1(x)}{[1 - A(x)] W(x)}$

We are thus led to the *integro-differential* equation

$$y(x) = \gamma_1 F_1(x) + \gamma_2 F_2(x) + \int_{x_0}^{x} L(x, \xi) [M(\xi) y(\xi) + N(\xi) y'(\xi)] \, d\xi$$

which reduces immediately to an ordinary integral equation on integration by parts; for since $L(x, x) \equiv 0$

$$\int_{x_0}^{x} L(x, \xi) N(\xi) y'(\xi) \, d\xi = -L(x, x_0) N(x_0) y(x_0) - \int_{x_0}^{x} \frac{\partial}{\partial \xi} [L(x, \xi) N(\xi)] y(\xi) \, d\xi$$

and the integral equation of the problem assumes the form

(25) $\quad y(x) - \displaystyle\int_{x_0}^{x} \left\{ L(x, \xi) M(\xi) - \frac{\partial}{\partial \xi} [L(x, \xi) N(\xi)] \right\} y(\xi) \, d\xi$

$$= \gamma_1 F_1(x) + \gamma_2 F_2(x) - L(x, x_0) N(x_0) y(x_0)$$

In view of (13) and of the significance of γ_1 and γ_2, this right-hand side may also be written as

$$\gamma_1[F_1(x) - L(x, x_0) N(x_0) F_1(x_0)] + \gamma_2[F_2(x) - L(x, x_0) N(x_0) F_2(x_0)]$$

In the particular case treated by Liouville-Stekloff in which $A(x) \equiv B(x) \equiv N(x) \equiv 0$ the nucleus of the integral equation (25) is simply equal to

$$(26) \qquad K'(x, \xi) = L(x, \xi) \frac{C(\xi)}{W(\xi)}$$

and its right-hand side equals $\gamma_1 F_1(x) + \gamma_2 F_2(x)$.

By any of these reductions the problem is reduced to the solution of one or more *linear integral equations of the second kind of Volterra type** which, on writing for convenience $x_0 = 0$ and (for reasons analogous to those considered in the Fredholm case) introducing a parameter multiplying the integral,** may be written in the standardized form

$$(27) \qquad \phi(x) - \lambda \int_0^x K(x, y) \phi(y) \, dy = f(x)$$

Here the *nucleus* K is not necessarily of either form (20) or (26).

Equations of type (27) can be solved fairly simply by the method of successive approximations outlined in § 3; this process is considerably simplified in this case due to the linearity of the integrand in the unknown function ϕ. Writing***

$$\phi(x) = f(x) + \sum_{n=1}^{\infty} \lambda^n \psi_n(x)$$

and collecting all the terms on one side of the equation, and successively equating to zero the coefficients of λ^n, we obtain the recurrence relations

$$\psi_n(x) = \int_0^x K(x, y) \psi_{n-1}(y) \, dy \qquad (n = 0, 1, 2, \ldots; \quad \psi_0 \equiv f)$$

which lead immediately to the general formula

$$\psi_n(x) = \int_0^x K_n(x, y) f(y) \, dy \qquad (n = 1, 2, \ldots)$$

---

* Compare with (79) of § 27, which is an equation of Fredholm type.
** In the above case $\lambda = 1$.
*** For greater detail see F. Tricomi (82), Chapter I.

where $K_1 \equiv K, K_2, K_3, \ldots$, are the successive *iterated nuclei* of K defined by the recurrence relations

$$(28) \qquad K_{n+1}(x, y) = \int_y^x K(x, z) K_n(z, y) \, dz \qquad (n = 1, 2, \ldots)$$

If now we define a *resolvent nucleus* H by the series

$$(29) \qquad H(x, y; \lambda) = - \sum_{n=0}^{\infty} \lambda^n K_{n+1}(x, y)$$

we obtain

$$(30) \qquad \phi(x) = f(x) - \lambda \int_0^x H(x, y; \lambda) f(y) \, dy$$

With regard to the convergence of the method nothing need be added to what has been said in Chapter I as a linear function fully satisfies the Lipschitz condition. In any case it follows immediately from (28) that if $|K(x, y)| < N$ then*

$$|K_{n+1}(x, y)| < N^{n+1} \frac{(x-y)^n}{n!} \leq N \frac{(hN)^n}{n!} \qquad (n = 1, 2, \ldots; \ 0 \leq y \leq x \leq h)$$

thus ensuring the absolute and uniform convergence of the series (29) which is the essential thing.

Substituting the appropriate expression of type (30) for the solution of the Volterra equation (19) into the preceding formula (23) we obtain a fundamental system of integrals of the given equation (10) in the explicit form

$$(31) \qquad Y_h(x) = F_h(x) + \int_{x_0}^x L(x, \xi) \mathcal{G}_h(\xi) \, d\xi$$
$$+ \sum_{n=1}^{\infty} \int_{x_0}^x L(x, \xi) \, d\xi \int_{x_0}^{\xi} K_n(\xi, \eta) \mathcal{G}_h(\eta) \, d\eta \qquad (h = 1, 2)$$

where K_n denotes the nth iterated nucleus of the nucleus K given by (20) and where L is given by (21).

The corresponding result in the case treated by Liouville-Stekloff is simpler; there

$$(32) \qquad Y_h(x) = F_h(x) + \sum_{n=1}^{\infty} \int_{x_0}^x K_n'(x, \xi) F_h(\xi) \, d\xi \qquad (h = 1, 2)$$

---

* For suitable but less restrictive conditions on K see the following section, and also F. TRICOMI (82).

where K_1', K_2', K_3', ..., are successive iterates of the nucleus (26); (32) may also be written as

$$(32') \qquad Y_h(x) = F_h(x) - \int_{x_0}^{x} H'(x, \xi) F_h(\xi) \, d\xi$$

where $H'(x, \xi)$ denotes the resolvent nucleus in the case $\lambda = 1$, i.e. where

$$(33) \qquad H'(x, \xi) = -\sum_{n=0}^{\infty} K'_{n+1}(x, \xi)$$

The importance of the method outlined above lies mainly in the fact (as is easily seen) that it does not demand that the coefficients $A(x)$, $B(x)$, $C(x)$ on the right-hand side of (10) should be 'small'; it is sufficient that they be finite. It is however easily seen that the method is particularly effective when A, B, C—and therefore also $\mathscr{G}_1$, $\mathscr{G}_2$ and the nucleus K given by (20) and all its iterates—are effectively 'small'.

For example, if the given equation contains a certain parameter ν, which is to tend to infinity, and the three coefficients $A(x)$, $B(x)$, $C(x)$ are (uniformly with respect to x) infinitesimal of a certain order $r > 0$ with respect to ν, i.e. if

$$(34) \qquad A(x) = O(\nu^{-r}), \qquad B(x) = O(\nu^{-r}), \qquad C(x) = O(\nu^{-r})$$

we have

$$\mathscr{G}_h(x) = O(\nu^{-r}), \qquad K_1(x, \xi) = O(\nu^{-r}), \qquad K_2(x, \xi) = O(\nu^{-2r}), \ldots$$

Therefore from (31) follow the asymptotic formulæ

$$(35) \qquad Y_h(x) = F_h(x) + O(\nu^{-r}) \qquad\qquad (h = 1, 2)$$

which are frequently extremely useful. If these results are not sufficiently sharp we may write

$$(35') \qquad Y_h(x) = F_h(x) + \int_{x_0}^{x} L(x, \xi) \mathscr{G}_h(\xi) \, d\xi + O(\nu^{-2r}) \qquad (h = 1, 2)$$

and so on.

30. Differential equations with stable integrals

In many questions in mechanics and physics it is of considerable importance to determine whether all (or some) of the integrals of a given linear differential equation of second order are *stable* or not, i.e. whether or not they remain bounded as the independent variable tends to $+\infty$ or $-\infty$.

§ 30]   *Differential equations with stable integrals*

We suppose for simplicity that the equation has been reduced to the form (18) of Chapter III; we start by considering the following question:

Suppose that all the integrals of the equation

(36) $$y'' + Q_1(x) y = 0$$

are stable as $x \to +\infty$*[*]; *what are the conditions that must be imposed on the function* $Q_2(x)$ *to ensure that all the integrals of the equation*

(36') $$y'' + [Q_1(x) + Q_2(x)] y = 0$$

shall also be stable as $x \to +\infty$?

A well-known theorem of G. Ascoli[**] establishes that a *sufficient condition for this is that the function* $Q_2(x)$ *be absolutely integrable within the vicinity of the point* $+\infty$, *i.e. the integral*

(37) $$I = \int_{x_0}^{\infty} |Q_2(x)|\, dx$$

exists, where x_0 *is any fixed point.*

This theorem is an immediate corollary of an important bound for all the integrals of (36') which may be easily obtained by combining the method outlined in the last section with the following lemma on Volterra integral equations:

If the nucleus K of the Volterra equation (27) *of the second kind may be written in the form*

$$K(x, y) = K^*(x, y)\, k(y)$$

where $K^*(x, y)$ *denotes a function bounded within the triangle, finite or infinite,* $0 \le y \le x \le a$, *and* $k(y)$ *is an absolutely integrable function in* $0 \le y \le a$ *(*$k(y)$ *bounded or not) the integral equation* (27) *may be solved in the way outlined in the preceding section exactly as if K were bounded; also, given N an upper bound of* $|K^*(x, y)|$, *the solution* $\phi(x)$ *of the integral equation satisfies the inequality*

(38) $$|\phi(x) - f(x)| \le |\lambda| N J^*(x) e^{|\lambda| N J(x)} \qquad (0 \le x \le a)$$

---

[*] With suitable modifications if $x \to -\infty$.
[**] G. Ascoli: 'Sul comportamento asintotico delle equazioni differenziali del 2° ordine', *Rend. Lincei*, (6) **22**, 234–43 (1935). For later work of Tricomi and of others on this question, see G. Sansone (47); see also p. 158 and p. 161 of this book. Among the more recent literature are two interesting notes by G. Prodi: *Rend. Lincei*, (8) **10**, 447–51 and **11**, 30–34 (1951).

where

(39) $$J(x) = \int_0^x |k(y)|\, dy, \qquad J^*(x) = \int_0^x |k(y)f(y)|\, dy$$

We show first that under the above hypotheses the $(n+1)$th iterated nucleus K_{n+1} satisfies the condition

(40) $$|K_{n+1}(x, y)| \leq N|k(y)|\frac{[NJ(x)]^n}{n!} \qquad (n = 0, 1, 2, \ldots)$$

Obviously (40) is valid for $n = 0$; now assuming the result valid for $n-1$ we show its validity for n. Since

$$\frac{dJ(x)}{dx} = |k(x)|$$

we have

$$|K_{n+1}(x, y)| \leq \int_y^x |k(z)K^*(x, z)||K_n(z, y)|dz \leq \frac{N^{n+1}}{(n-1)!}|k(y)|\int_y^x |k(z)|[J(z)]^{n-1}dz$$

$$= \frac{N^{n+1}}{(n-1)!}|k(y)|\int_y^x [J(z)]^{n-1}\frac{dJ(z)}{dz}dz = \frac{N^{n+1}}{(n-1)!}|k(y)|\frac{1}{n}[\{J(z)\}^n]_y^x$$

$$= \frac{N^{n+1}}{n!}|k(y)|[\{J(x)\}^n - \{J(y)\}^n] \leq N|k(y)|\frac{[NJ(x)]^n}{n!}$$

as was to be proved.

From (40) it follows immediately that the series (29) which gives the resolvent nucleus $H(x, y; \lambda)$ admits as majorant the series

$$N|k(y)|\sum_{n=0}^{\infty}\frac{[|\lambda|NJ(x)]^n}{n!} = N|k(y)|e^{|\lambda|NJ(x)}$$

Hence (everywhere $k(y)$ is finite) the series (29) is itself convergent (absolutely and almost uniformly) and

$$|H(x, y; \lambda)| \leq N|k(y)|e^{|\lambda|NJ(x)}$$

Now using (30) we obtain the inequality

$$|\phi(x) - f(x)| \leq |\lambda|\int_0^x Ne^{|\lambda|NJ(x)}|k(y)f(y)|\, dy$$

which is in fact (38).

We now apply the method of the preceding section to (36') written in the form

$$y'' + Q_1(x)y = -Q_2(x)y$$

This is equivalent to saying that with two linearly independent integrals $F_1(x)$ and $F_2(x)$ of (36) (with Wronskian equal to 1) may be associated two similar integrals $Y_1(x)$ and $Y_2(x)$ of (36') such that

$$Y_h(x) = F_h(x) - \int_{x_0}^{x} H(x, \xi) F_h(\xi) d\xi \qquad (h = 1, 2)$$

where x_0 denotes some fixed abscissa and $H(x, \xi)$ the resolvent nucleus corresponding (in the case $\lambda = 1$) to the nucleus

$$K(x, \xi) = L(x, \xi) C(\xi) = [F_1(x) F_2(\xi) - F_2(x) F_1(\xi)] Q_2(\xi)$$

But equation (36) has, by hypothesis, all its integrals *stable*; thus there exist two positive constants A_1 and A_2 such that for all $x \geq x_0$ (and for $x \to +\infty$)

(41) $$|F_1(x)| \leq A_1, \qquad |F_2(x)| \leq A_2$$

and therefore

$$|F_1(x) F_2(\xi) - F_2(x) F_1(\xi)| \leq 2 A_1 A_2$$

Hence using the above lemma and putting

(42) $$\int_{x_0}^{x} |Q_2(y)| dy = J(x)$$

we obtain

$$J_h^*(x) = \int_{x_0}^{x} |Q_2(y) F_h(y)| dy \leq A_h J(x) \qquad (h = 1, 2)$$

and therefore

(43) $$|Y_h(x) - F_h(x)| \leq 2 A_1 A_2 A_h J(x) e^{2 A_1 A_2 J(x)} \qquad (h = 1, 2)$$

from which follow

(43') $$|Y_h(x)| \leq A_h \{1 + 2 A_1 A_2 J(x) e^{2 A_1 A_2 J(x)}\} \qquad (h = 1, 2)$$

From this last inequality we can immediately deduce the theorem of Ascoli referred to earlier. If the integral (37) exists, we have $J(x) \leq I$ for all x, and from (43') follow

(43'') $$|Y_h(x)| \leq A_h \{1 + 2 A_1 A_2 I e^{2 A_1 A_2 I}\} \qquad (h = 1, 2)$$

which shows that the integrals of (36') are also stable.

The preceding inequalities contain much more than the theorem just established. For example, if $J(x)$ is not only bounded as $x \to +\infty$ but is also of a lower order of magnitude than that of $F_h(x)$, from (43) we may immediately deduce the asymptotic formula

$$(44) \qquad Y_h(x) = F_h(x) + O[J(x)] \qquad (x \to +\infty)$$

which is of considerable importance. Writing this result in different form, we may assert that with any integral $Y(x)$ of (36′) may be associated two constants γ_1 and γ_2 such that as $x \to +\infty$

$$(44') \qquad Y(x) = \gamma_1 F_1(x) + \gamma_2 F_2(x) + O[J(x)]$$

We now consider an important application of these results, and for this purpose we observe that one equation of type (36) all of whose integrals are stable is the equation of the harmonic oscillator (§ 26)

$$(45) \qquad y'' + k^2 y = 0$$

where k^2 is any positive constant. The following important corollary is immediately deducible from Ascoli's theorem.

If the coefficient $Q(x)$ in the differential equation

$$(46) \qquad y'' + Q(x) y = 0$$

tends to a positive limit k^2 as $x \to +\infty$ and if the integral

$$\int_{x_0}^{\infty} |Q(x) - k^2|\, dx$$

exists where x_0 is any fixed abscissa, all the integrals of equation (46) are stable as $x \to +\infty$.

We now apply this result to the equation

$$(47) \qquad \frac{d^2 y}{dx^2} + \left(1 + \frac{\frac{1}{4} - v^2}{x^2}\right) y = 0$$

which, as was shown in § 21, is obtained from the *Bessel equation* by multiplying its dependent variable by $\sqrt{x}$.

In this example $k = 1$, and two linearly independent integrals of the 'approximate' equation (with Wronskian equal to 1) are $\cos x$ and $\sin x$; also, if x_0 is any positive quantity,

$$J(x) = |\tfrac{1}{4} - v^2| \int_{x_0}^{x} \frac{d\xi}{\xi^2} = |\tfrac{1}{4} - v^2|\left(\frac{1}{x_0} - \frac{1}{x}\right) < \frac{|\tfrac{1}{4} - v^2|}{x_0}$$

Therefore if $Z_\nu(x)$ is any Bessel function of order ν, i.e. if $\sqrt{x}Z_\nu(x)$ is any solution of (47), there exist two constants C_ν and γ_ν such that

$$\sqrt{x}\,Z_\nu(x) = C_\nu[\sin\gamma_\nu\, Y_1(x) + \cos\gamma_\nu\, Y_2(x)]$$

where $Y_1(x)$ and $Y_2(x)$ are solutions of (47) satisfying, in view of (43) (in which now $A_1 = A_2 = 1$), the inequalities

$$|Y_1(x) - \cos x| \leq 2J(x)e^{2J(x)}, \qquad |Y_2(x) - \sin x| \leq 2J(x)e^{2J(x)}$$

from which we deduce immediately that

$$|\sqrt{x}\,Z_\nu(x) - C_\nu \sin(x+\gamma_\nu)| \leq 4C_\nu J(x)e^{2J(x)}$$

Now x_0 is any positive number, and we may therefore suppose $x_0 \geq 1$; thus

$$e^{2J(x)} < e^{2|\frac{1}{4}-\nu^2|/x_0} \leq e^{2|\frac{1}{4}-\nu^2|} = A$$

and the preceding inequality may be written in the simpler form

(48) $$|\sqrt{x}\,Z_\nu(x) - C_\nu \sin(x+\gamma_\nu)| < AC_\nu \frac{|1-4\nu^2|}{x_0} \qquad (x \geq x_0 \geq 1)$$

Further, since x_0 is arbitrarily chosen this implies that the quantity on the left-hand side (which does not depend on x_0) is $O(x^{-1})$ as $x \to +\infty$. Therefore, dividing by $\sqrt{x}$, we obtain the important asymptotic formula

(49) $$Z_\nu(x) = \frac{C_\nu}{\sqrt{x}} \sin(x+\gamma_\nu) + O(x^{-3/2})$$

For example, in the case of the *Bessel function of the first kind* $J_\nu(x)$ which may be represented by the series, everywhere convergent,

$$J_\nu(x) = \sum_{n=0}^{\infty} \frac{(-1)^n}{n!\,\Gamma(n+\nu+1)}\left(\frac{x}{2}\right)^{2n+\nu}$$

where Γ is the symbol for the *Gamma function**, it can be shown that

$$C_\nu = \sqrt{\frac{2}{\pi}}, \qquad \gamma_\nu = -\frac{\nu\pi}{2} + \frac{\pi}{4}$$

---

* For a brief note on this function see F. TRICOMI (79); more information is to be found in W. MAGNUS-F. OBERHETTINGER (69).

thus giving the result

(50) $$J_\nu(x) = \sqrt{\frac{2}{\pi x}} \, \sin\left(x - \frac{\nu\pi}{2} + \frac{\pi}{4}\right) + O(x^{-3/2})$$

The determination of these two constants however is not easy.*

These formulæ clearly illustrate the *damped oscillatory* behaviour of the Bessel functions, and also the fact that the distance between two successive zeros tends to π as $|x|$ increases.

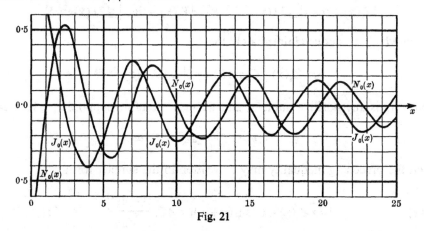

Fig. 21

In figure 21 are shown the graph of the function $J_0(x)$ in the interval from $x = 0$ to $x = 25$, and that of the function $N_0(x)$, a solution of the Bessel equation linearly independent of $J_0(x)$ and called a function of *the second kind*. In the formula (49) corresponding to the function $N_\nu(x)$ the constant C_ν has the same value $\sqrt{(2/\pi)}$ as for J_ν, but the value of γ_ν is increased by $\pi/2$, i.e. for $N_\nu(x)$ we require $\gamma_\nu = -\nu\pi/2 + 3\pi/4$**.

31. The case in which the coefficient of y tends to a negative limit

When the coefficient $Q(x)$ in (46) tends to a *negative* limit $-k^2$ as $x \to +\infty$ most of the formulæ of the last section, perhaps with obvious modifications, still remain valid but they lack interest since only one of the two integrals

$$F_1(x) = e^{kx}, \qquad F_2(x) = e^{-kx}$$

---

* For the case in which ν is integral, see, for example, G. SANSONE (47), Vol. II, pp. 38–42.

** The most comprehensive work on Bessel functions is G. N. WATSON, *A Treatise on Bessel Functions* (2nd edition, Cambridge University Press, 1945). In the *Funktionentafeln* by E. JAHNKE-F. EMDE (66) the part relative to Bessel functions is excellent.

of the 'approximate' equation

$$y'' - k^2 y = 0$$

is stable as $x \to +\infty$.

We can also obtain bounds of the type (43) but at the cost of imposing drastic conditions on $[Q(x)-(-k^2)]$ in order to preserve the convergence of the improper integrals whose integrands contain this difference function multiplied by functions which tend exponentially to infinity as $x \to \infty$.*

It is more convenient however to adopt a different method and we consider, as has already been done in § 19, suitable quadratic forms in y and y'. In this way we easily establish the following important theorem which is due initially to Poincaré.**

If as $x \to +\infty$ the coefficient $Q(x)$ of y in equation (46) tends to a NEGATIVE *limit which for simplicity we suppose to be -1, then there exist integrals of equation (46) which tend to infinity as $x \to +\infty$ in such a way that*

(51) $$\lim_{x \to \infty} \frac{y'}{y} = 1$$

while all other integrals tend to ZERO *in such a way that*

(51') $$\lim_{x \to \infty} \frac{y'}{y} = -1$$

To prove this result we consider the quadratic form

$$v(x) = y(x)\,y'(x) = \tfrac{1}{2} \frac{d}{dx} y^2(x)$$

(already considered in § 19); thus

(52) $$v'(x) = y'^2 + yy'' = y'^2 - Q(x)\,y^2$$

and this derivative is certainly positive to the right of a suitable x_0 since $Q(x)$ (which tends to -1) must necessarily be of negative sign from some point onwards. It consequently follows that $v(x)$ *is always increasing for* $x > x_0$ and therefore only two cases can arise:

(1) There exists an $x_1 \geqq x_0$ such that $v(x_1) \geqq 0$, and therefore for $x > x_1$, $v(x) > 0$.

(2) The function $v(x)$, while increasing, always remains negative to the right of x_0.

---

* See G. SANSONE (47), Vol. II, p. 51.
** H. POINCARÉ: 'Sur les équations linéaires aux différentielles ordinaires et aux différences finies', *Amer. J. of Math.*, 7, 203–58 (1885).

In both cases $y(x)$ must ultimately behave *monotonically* since neither y' nor y can vanish when $v(x) \neq 0$; hence in the first case $|y|$ will be *increasing*, and in the second case will be instead *decreasing*. We may in fact speak of y rather than $|y|$ as there is no essential restriction in supposing that y, which cannot vanish to the right of x_1 or of x_0, is always *positive*—as the integrals of a homogeneous equation such as (46) may always be multiplied by any constant factor.

Also $y'(x)$ tends to a certain limit, finite or infinite, as $x \to +\infty$, since on multiplying (46) by $2y'$ we have

$$\frac{d}{dx} y'^2 = -2Q(x)\,v(x)$$

thus showing that this derivative also is ultimately of constant sign (positive in the first of the two cases and negative in the second) so that $y'(x)$ from some point onwards behaves monotonically.

We now consider the first of the two above cases, and denote by λ the limit, infinite or finite but certainly not zero, to which y tends as $x \to +\infty$, and by μ the analogous limit to which the function $v(x)$ tends. On the one hand we have

$$\lim_{x \to \infty} y'(x) = \lim_{x \to \infty} \frac{v(x)}{y(x)} = \frac{\mu}{\lambda}$$

while on the other, by l'Hospital's rule,* we have

$$\lim_{x \to \infty} \frac{y}{x} = \lim_{x \to \infty} y'(x)$$

This implies that *the limit to which y tends cannot be finite* since then $\lim (y/x) = 0$ and therefore $\lim y'(x) = 0$, which contradicts the first of the two preceding limit relations, as when $\mu \neq 0$, μ/λ cannot be zero when λ is finite.

In addition, again applying l'Hospital's rule, we have

(53) $$\lim_{x \to \infty} \frac{y'^2}{y^2} = \lim_{x \to \infty} \frac{2y'y''}{2y'y} = -\lim_{x \to \infty} Q(x) = 1$$

from which, as $v(x) = yy' > 0$, we immediately deduce (51).

We proceed to the second case: λ again denotes the limit, now certainly finite and non-negative, to which y tends as $x \to +\infty$.

---

* Here and in the following sections we shall use l'Hospital's rule in the form given by Stolz; this states that *if the limit of u'/v' exists and if the function v tends to infinity, then the limit of u/v also exists and equals the limit of u'/v'*. NOTE no assumption is made about the limit of u. A proof of this may be found in F. TRICOMI (79), Vol. II, Chapter VII, § 4.

§ 31]   The case in which the coefficient of y tends to a negative limit   157

In this case λ must be zero; for if λ were not zero it would follow from (52) that

$$\lim_{x \to \infty} v'(x) = \lim_{x \to \infty} \frac{v^2(x)}{y^2(x)} - \lim_{x \to \infty} [Q(x)y^2] = \frac{\mu^2}{\lambda^2} + \lambda^2 > 0$$

in contradiction of the fact that (again by l'Hospital's rule) this limit must coincide with that of $v(x)/x$ which is certainly zero, as the limit of $v(x)$ as $x \to +\infty$ is finite.

Further, we see that y' must also tend to *zero* as $x \to +\infty$; for, again applying l'Hospital's rule, this limit must coincide with the limit of y/x which is zero.

Since both y and y' tend to *zero* we can make use of (53) to derive the required result, for in this case y and y' are of opposite sign as $v(x) = yy' < 0$, and (51') follows.

We now show that there is always at least one integral y_1 of (46) for which the first of the two cases occurs and an integral y_2 for which the second case occurs.

For y_1 the result is immediate as it is sufficient to consider an integral defined by the initial conditions

$$y(x_0) = \alpha, \qquad y'(x_0) = \alpha'$$

where α and α' are of the same sign, to ensure that $v(x) = yy'$ is positive at $x = x_0$ (and therefore to the right of this point).

It follows from (51), which may also be written in the form

$$\frac{d}{dx} \log y_1 = 1 + o(1)$$

that the integral y_1 may be represented asymptotically by

(54) $$y_1(x) = C_1 e^{x[1 + \varepsilon(x)]}$$

where $\varepsilon(x)$ denotes a function which tends to *zero* as $x \to +\infty$ and C_1 a suitable constant.*

Using this result together with the formula of Liouville,** it follows that

---

* We are using the result that if a function $\varphi(x)$ tends to *zero* as $x \to \infty$, then

$$\int_0^x \varphi(t) dt = o(x)$$

This follows by applying l'Hospital's rule to obtain the limit of the ratio of this integral to x—this limit is the limit of $\varphi(x)$.

** See, for example, F. TRICOMI (79), Part II, p. 301.

if y_2 be any chosen integral of (46) linearly independent of y_1 we can write

$$y_1 y_2' - y_2 y_1' = k$$

i.e.

$$\frac{d}{dx}\left(\frac{y_2}{y_1}\right) = \frac{k}{y_1^2}$$

where k denotes a suitable non-zero constant. It is easily seen that at least the particular integral y_2 defined by the formula*

(55) $$\frac{y_2(x)}{y_1(x)} = -k \int_x^\infty \frac{dt}{y_1^2(t)}$$

belongs to the second of the two categories above.

As both the integral which appears in (55) and y_1^{-2} tend to zero as $x \to +\infty$ by l'Hospital's rule we again obtain

$$\lim_{x\to\infty} \frac{\int_x^\infty y_1^{-2}(t)\,dt}{y_1^{-2}(x)} = \lim_{x\to\infty} \frac{y_1^{-2}(x)}{2y_1^{-3}(x) y_1'(x)} = \tfrac{1}{2} \lim_{x\to\infty} \frac{y_1(x)}{y_1'(x)} = \tfrac{1}{2}$$

which evidently implies that $y_2 y_1$ tends to $-k/2$, and therefore

(56) $$y_2(x) = [-\tfrac{1}{2}k + o(1)] y_1^{-1}(x)$$

showing that $y_2(x)$ tends to zero with $y_1^{-1}(x)$ as $x \to \infty$.

Lastly we note that if y_1 and y_2 are two linearly independent integrals of equation (46) any other integral y is expressible in the form

$$y(x) = C_1 y_1(x) + C_2 y_2(x)$$

where C_1 and C_2 are two suitable constants; thus y is 'of the type y_1' or 'of the type y_2' according as $C_1 \neq 0$ or $C_1 = 0$.

The proof of the theorem enunciated on page 155 now being completed we propose to improve the asymptotic formula (54) by means of the following theorem whose proof is due to G. Ascoli.**

Suppose that $Q(x)$ tends to -1 as $x \to +\infty$ and let

(57) $$\begin{cases} Q(x) = -[1 + Q_2(x)] \\ \tau(x) = x + \tfrac{1}{2} \int_{x_0}^x Q_2(t)\,dt, \quad \sigma(x) = \int_{x_0}^x \sqrt{[1+Q_2(t)]}\,dt \end{cases}$$

Then every integral $y(x)$ of (46) is $O(e^\tau)$ as $x \to +\infty$; further, every integral which tends to infinity is of order not less than e^σ.

---

* From (54) it is easily seen that this integral is convergent.
** G. Ascoli: Boll. Unione Mat. Ital., (3) 8, 115–23 (1953).

§ 31] The case in which the coefficient of y tends to a negative limit

If the function Q_2 satisfies the further condition

(58) $$\int_{x_0}^{\infty} Q_2^2(x)\,dx < \infty$$

the difference $\tau - \sigma$ tends to a finite limit as $x \to +\infty$; also, for every integral $y_1(x)$ tending to infinity

(59) $$y_1(x) = [h + o(1)]e^{\tau(x)}$$

while for every integral $y_2(x)$ tending to zero

(59′) $$y_2(x) = [h' + o(1)]e^{-\tau(x)}$$

where h and h' are two non-zero constants.*

(It is obvious that the fixed abscissa x_0 in the integral σ must be supposed sufficiently large that $1 + Q_2 > 0$.)

For proof consider the quadratic form in y and y'

$$u(x) = y^2 + y'^2$$

from which, on differentiating $u(x)$ and making use of the differential equation (46), we obtain

$$u'(x) = 2yy' + 2y'y'' = 4yy'[1 + \tfrac{1}{2}Q_2(x)]$$

Since for any two real numbers a and b

(60) $$2ab \leq a^2 + b^2$$

in particular

$$2yy' \leq y^2 + y'^2 = u(x)$$

Thus supposing that x is sufficiently large that

$$1 + \tfrac{1}{2}Q_2(x) > 0$$

we may write

$$u'(x) \leq 2u(x)[1 + \tfrac{1}{2}Q_2(x)]$$

and on dividing by u (which is positive) we obtain the inequality

$$\frac{d}{dx}[\log u(x) - 2\tau(x)] = \frac{d}{dx}\log[e^{-2\tau(x)}u(x)] \leq 0$$

---

* The asymptotic formulæ (59), (59′) are equivalent to those obtained under the same condition (58) but by more complicated methods by PH. HARTMAN: 'Unrestricted solution fields of almost-separable differential equations', *Trans. Amer. Math. Soc.*, 63, 560–80 (1948).

It follows from this that at least from a certain point onwards the positive function

$$e^{-2\tau(x)}u(x)$$

is *non-increasing* and therefore as $x \to +\infty$ will tend to a certain finite limit, *positive or zero*, which we denote by $2h^2$; thus

$$2h^2 = \lim_{x \to \infty} e^{-2\tau}(y^2 + y'^2) = \lim_{x \to \infty} e^{-2\tau}y^2\left(1 + \frac{y'^2}{y^2}\right) = \lim_{x \to \infty} e^{-2\tau}y'^2\left(1 + \frac{y^2}{y'^2}\right)$$

But by (53)

$$\lim_{x \to \infty}\left(1 + \frac{y'^2}{y^2}\right) = \lim_{x \to \infty}\left(1 + \frac{y^2}{y'^2}\right) = 2$$

and it follows that the two functions

$$e^{-2\tau}y^2, \quad e^{-2\tau}y'^2$$

tend to the common limit h^2; we therefore deduce immediately that when h is of suitable sign

(61) $$\lim_{x \to \infty} e^{-\tau}y = h, \quad \lim_{x \to \infty} e^{-\tau}y' = \pm h$$

It follows that not only y but also y' is at most (*at most* since h might be zero) of the order of e^{τ}, as $x \to +\infty$.

Now consider again the quadratic form $v(x) = yy'$; using the inequality (60) with

$$a = y', \quad b = \sqrt{[1 + Q_2(x)]}\, y$$

we have

$$v'(x) = y'^2 + [1 + Q_2(x)]y^2 \geq 2yy'\sqrt{[1 + Q_2(x)]} = 2v(x)\sigma'(x)$$

Dividing throughout by $v(x)$ which we shall suppose positive* we may write

$$\frac{v'(x)}{v(x)} - 2\sigma'(x) = \frac{d}{dx}\log[e^{-2\sigma(x)}v(x)] \geq 0$$

and therefore the positive function

$$e^{-2\sigma(x)}v(x)$$

is, at least from a certain point onwards, *non-decreasing*. This function

---

* This implies that the first of the two cases previously discussed is appropriate here, i.e. that y tends to infinity as $x \to +\infty$.

therefore tends to a limit, finite or infinite but certainly non-zero, which we denote by l, i.e.

$$\lim_{x\to\infty} e^{-2\sigma}v(x) = \lim_{x\to\infty} e^{-\sigma}y \cdot e^{-\sigma}y' = l > 0$$

and this would not be so were y and y', which are of the same order as $x \to +\infty$ (since their ratio tends to 1), of order less than e^σ.

This establishes the first part of the theorem. To complete the proof we use the identity

$$\tau' - \sigma' = 1 + \tfrac{1}{2}Q_2 - \sqrt{[1+Q_2]} = \frac{Q_2^2}{4(1+\tfrac{1}{2}Q_2+\sqrt{[1+Q_2]})}$$

from which follows

(62) $$\tau(x) - \sigma(x) = x_0 + \tfrac{1}{4}\int_{x_0}^{x} \frac{Q_2^2(t)\,dt}{1+\tfrac{1}{2}Q_2(t)+\sqrt{[1+Q_2(t)]}}$$

Now if we suppose x_0 sufficiently large that for $t \geq x_0$, $|Q_2(t)| < 1/2$, then

$$1 + \tfrac{1}{2}Q_2(t) + \sqrt{[1+Q_2(t)]} > 1 - \tfrac{1}{4} + \sqrt{[1-\tfrac{1}{2}]} > \tfrac{3}{4}$$

from which follows

$$\tfrac{1}{4}\left|\int_{x_0}^{x} \frac{Q_2^2(t)\,dt}{1+\tfrac{1}{2}Q_2(t)+\sqrt{[1+Q_2(t)]}}\right| < \tfrac{1}{3}\int_{x_0}^{x} Q_2^2(t)\,dt < \tfrac{1}{3}\int_{x_0}^{\infty} Q_2^2(t)\,dt$$

Thus provided (58) is valid the difference $\tau(x) - \sigma(x)$ remains bounded as $x \to +\infty$ and the same is therefore true for the ratio e^τ/e^σ.

We therefore conclude that when the condition (58) is satisfied formula (59) *with h a non-zero constant* is valid for every integral y_1 tending to infinity as $x \to +\infty$, or, instead, a similar formula in which $\sigma(x)$ appears in place of $\tau(x)$.

To complete the proof we note that it follows immediately from (56) that if (59) with $h \neq 0$ is valid for every integral 'of type y_1' then (59') is valid for every integral 'of type y_2'.

Other asymptotic representations in which an additional term

$$-\tfrac{1}{2}\int_0^x Q_2(t_1)\,dt_1 \int_{t_1}^x e^{2t_1 - 2t_2} Q_2(t_2)\,dt_2$$

is added to the integral $\tau(x)$ have recently been obtained by R. Bellman*

---

* R. BELLMAN: 'On the asymptotic behaviour, etc.', *Ann. di Mat.* (4) **31**, 83–91 (1950). This paper contains a minor error. See also G. ASCOLI: 'Sul comportamento asintotico degli integrali dell'equazione $y'' = [1 + f(t)]y$ in un caso notevole', *Riv. Mat. Univ. Parma*, **4**, 11–29 (1953).

under the less restrictive hypothesis that (58) be replaced by a similar integral containing $|Q_2(x)|^3$.

In the note by Ascoli cited earlier it is proved that if the function $Q_2(t)$ instead of satisfying the condition (58) is assumed to be *of bounded variation** in the interval $(x_0, +\infty)$ the formulæ (59), (59′) remain valid (with non-zero values of the constants h and h') *provided that in these formulæ $\tau(x)$ is now replaced by the integral $\sigma(x)$*, as it is now no longer certain that these two integrals are of the same order as $x \to +\infty$.

We shall not consider the cases in which $Q(x)$ tends to *zero* or to *infinity* as $x \to \infty$,** other than to mention as illustrative example of an equation in which $Q(x)$ tends to zero, that derived from the function

$$y = x^\alpha \sin(x^\beta + c)$$

for which

$$y' = \alpha x^{\alpha-1} \sin(x^\beta + c) + \beta x^{\alpha+\beta-1} \cos(x^\beta + c)$$

$$y'' = \left(\frac{\alpha(\alpha-1)}{x^2} - \frac{\beta^2}{x^{2-2\beta}}\right) x^\alpha \sin(x^\beta + c) + \beta(\beta + 2\alpha - 1) x^{\alpha+\beta-2} \cos(x^\beta + c)$$

If α and β are related by the equation

$$\beta(\beta + 2\alpha - 1) = 0$$

this function satisfies the differential equation of type (46)

(63) $$y'' + \left(\frac{\beta^2}{x^{2-2\beta}} + \frac{\alpha(1-\alpha)}{x^2}\right) y = 0$$

The example is interesting since if $0 < \beta < 1$, say $\beta = 1/2$ and therefore $\alpha = 1/4$, the integral from which we started is oscillatory as $x \to +\infty$; on the other hand, since

$$\lim_{x \to \infty} \left(\frac{\beta^2}{x^{2-2\beta}} + \frac{\alpha(1-\alpha)}{x^2}\right) = 0$$

the limit equation corresponding to (63) is simply $y'' = 0$ with general integral $y = c_1 x + c_2$, in which no oscillatory part appears. Equation (63) therefore provides a good example of the fact to which reference has already been made, that the asymptotic behaviour of the integrals of a differential equation cannot be deduced in all cases from the behaviour of the integrals of the limit equation.

---

* See later (§ 33) a note on functions of bounded variation.
** For a discussion of these cases see G. SANSONE (47), Chapter VII.

32. Preliminaries to the asymptotic treatment of eigenvalues and of eigenfunctions

This and subsequent sections contain a deeper discussion by asymptotic methods of the eigenvalues and of the eigenfunctions of a Sturm-Liouville system for large values of the index n, and extend the results of § 26.

It is sometimes convenient to modify the calculations of § 18 which reduced the given equation to another containing only a term in y'' and a term in y, so that the coefficient of y in the equation—which we have throughout supposed to be a *linear* function of the parameter λ—reduces to the form $q(x)+\lambda$, i.e. that the function previously denoted by $r(x)$ reduces to *unity*. The usefulness of this transformation has already been apparent in § 26.

We consider first the reduction of the second-order general homogeneous linear equation

$$A(x)\frac{d^2y}{dx^2}+B(x)\frac{dy}{dx}+C(x)y = 0$$

to a form in which the coefficient of the first derivative is zero.

Let

$$y = \alpha(x)z$$

where $\alpha(x)$ is a given function and z the new unknown variable; this leads to an equation similar to the original, viz.

$$A_1(x)\frac{d^2z}{dx^2}+B_1(x)\frac{dz}{dx}+C_1(x)z = 0$$

in which

$$A_1 = A\alpha, \qquad B_1 = 2A\alpha'+B\alpha, \qquad C_1 = A\alpha''+B\alpha'+C\alpha$$

Now make the change of independent variable

$$\xi = \int \phi(x)\,dx$$

which implies

$$\frac{dz}{dx} = \phi(x)\frac{dz}{d\xi}, \qquad \frac{d^2z}{dx^2} = \phi^2(x)\frac{d^2z}{d\xi^2}+\phi'(x)\frac{dz}{d\xi}$$

The given equation becomes

$$A_1\phi^2\frac{d^2z}{d\xi^2}+(A_1\phi'+B_1\phi)\frac{dz}{d\xi}+C_1 z = 0$$

and the coefficient of $dz/d\xi$ is zero provided

$$A_1\phi' + B_1\phi = 0$$

i.e. provided that the function ϕ is chosen so that

$$\frac{\phi'}{\phi} = -\frac{B_1}{A_1} = -\frac{2A\alpha' + B\alpha}{A\alpha} = -2\frac{\alpha'}{\alpha} - \frac{B}{A}$$

whence, by integration,

$$\phi(x) = \frac{1}{\alpha^2(x)} e^{-\int \frac{B(x)}{A(x)} dx}$$

We now apply this transformation to the equation

(64) $$\frac{d}{dx}\left[p(x)\frac{dy}{dx}\right] + [q(x) + \lambda r(x)] y = 0$$

in which

$$A(x) = p(x), \quad B(x) = p'(x), \quad C(x) = q(x) + \lambda r(x), \quad \phi(x) = \frac{1}{\alpha^2(x) p(x)}$$

so that

(65) $$y = \alpha(x) z, \quad \xi = \int \frac{dx}{\alpha^2(x) p(x)}$$

The equation then becomes

$$\frac{1}{\alpha^3 p}\frac{d^2 z}{d\xi^2} + [p\alpha'' + p'\alpha' + (q + \lambda r)\alpha] z = 0$$

and on dividing throughout by the leading coefficient and putting

$$q^*(\xi) = \alpha^3 p(p\alpha'' + p'\alpha' + q\alpha), \quad r^*(\xi) = \alpha^4 pr$$

we obtain the form

$$\frac{d^2 z}{d\xi^2} + [q^*(\xi) + \lambda r^*(\xi)] z = 0$$

or the simpler form

(66) $$\frac{d^2 z}{d\xi^2} + [q^*(\xi) + \lambda] z = 0$$

if the function α, which so far is undefined, is now chosen so that

$$r^*(\xi) = \alpha^4 pr = 1$$

§ 32]   *Preliminaries to the treatment of eigenvalues and eigenfunctions*

i.e.

(67) $$\alpha(x) = [p(x)\,r(x)]^{-\frac{1}{4}}$$

This choice for α is always possible within the real field if, as we shall suppose, $p(x) > 0$, $r(x) > 0$.

Thus *given an equation of type* (64) *in which* $p(x) > 0$, $r(x) > 0$ *and in which the product* $p(x)\,r(x)$ *may be twice differentiated, the transformations* (65) *in which* $\alpha(x)$ *has the value given in* (67), *viz.*

$$y = [p(x)\,r(x)]^{-\frac{1}{4}} z, \qquad \xi = \int \sqrt{\left(\frac{r(x)}{p(x)}\right)}\, dx$$

carry the given equation into equation (66) *where the value of* $q^*(\xi)$ *is that stated above.*

In studying the eigenvalues and the eigenfunctions of a Sturm-Liouville system we may therefore always refer to an equation of type (66); on replacing ξ by x, $q^*(\xi)$ by $-Q(x)$, and λ by μ^2,* we take as canonical form

(68) $$\frac{d^2 z}{dx^2} + [\mu^2 - Q(x)]\, z = 0$$

In addition, we may suppose that the fundamental interval be $(0, \pi)$ since if this is not the case initially we can make it so by a suitable change of independent variable.**

The simplification obtained by replacing (64) by (68) is illustrated by considering the fundamental inequality (72) of the preceding section; it now takes (for (68)) the simpler form

(69) $$(n-1)^2 + \min Q(x) < \mu_n^2 \leqq (n+1)^2 + \max Q(x)$$

(where $\lambda_n = \mu_n^2$) from which it follows immediately that $\mu_n^2 = n^2 + O(n)$, i.e. that

(69′) $$\mu_n = n + O(1)$$

The assumptions made above, that the functions $p(x)$ and $r(x)$ are both positive, are too restrictive in some cases that arise, and we shall shortly meet an example in which the function $r(x)$ may change sign.

---

* This substitution is justified since, as proved on page 119, when $r(x) > 0$ there can be only a finite number of negative eigenvalues; hence λ_n is certainly positive if n is sufficiently large.
From here onwards we suppose $\mu > 0$.

** The substitution $x' = \pi \dfrac{x-a}{b-a}$ transforms the interval $a \leqslant x \leqslant b$ into $0 \leqslant x' \leqslant \pi$.

If we suppose that the function $r(x)$ has a simple zero at the point x_0 and that the integral which defines ξ has x_0 as its lower limit (so that to $x = x_0$ corresponds $\xi = 0$) we may write

$$r(x) = \xi \rho(x)$$

where $\rho(x)$ is a function always non-zero in the vicinity of x_0 and which may be supposed always positive, if necessary by replacing λ by $-\lambda$; consequently we may replace (67) by

$$\alpha(x) = [p(x)\rho(x)]^{-\frac{1}{4}}$$

This substitution replaces (68) by

$$\frac{d^2 z}{dx^2} + [\lambda x - Q(x)] z = 0$$

in which (possibly by replacing x by $-x$) there is no restriction in supposing $\lambda > 0$; so that on putting $\lambda = \mu^2$ the equation assumes the canonical form

(68') $$\frac{d^2 z}{dx^2} + [\mu^2 x - Q(x)] z = 0$$

33. First form of asymptotic expression for the eigenfunctions

Since the 'approximate' equation $z'' + \mu^2 z = 0$ corresponding to (68) admits the fundamental system of integrals (with Wronskian equal to *one*)

$$F_1(x) = \frac{1}{\sqrt{\mu}} \cos \mu x, \qquad F_2(x) = \frac{1}{\sqrt{\mu}} \sin \mu x$$

on identifying (68) with (10), and therefore $C(x)$ with $Q(x)$, we immediately obtain from (32) two fundamental integrals of equation (68), expressed as absolutely and uniformly convergent series

$$\begin{cases} z_1(x) = \dfrac{1}{\sqrt{\mu}} \left\{ \cos \mu x + \sum_{n=1}^{\infty} \int_0^x K_n(x, \xi) \cos \mu \xi \, d\xi \right\} \\ z_2(x) = \dfrac{1}{\sqrt{\mu}} \left\{ \sin \mu x + \sum_{n=1}^{\infty} \int_0^x K_n(x, \xi) \sin \mu \xi \, d\xi \right\} \end{cases}$$

where $K_n(x, \xi)$ denotes the nth iterate of the nucleus

$$K(x, \xi) = [F_1(\xi) F_2(x) - F_2(\xi) F_1(x)] Q(\xi) = \frac{1}{\mu} \sin \mu (x - \xi) \cdot Q(\xi)$$

§ 33]   *First form of asymptotic expression for the eigenfunctions*   167

Further, if $A_1 \equiv A, A_2, A_3, \ldots$, are successive iterates of the nucleus

$$A(x, \xi) = \sin \mu(x-\xi) \cdot Q(\xi)$$

the general solution of (68) is

(70) $\qquad z(x) = (c_1 \cos \mu x + c_2 \sin \mu x)$

$$+ \sum_{n=1}^{\infty} \mu^{-n} \int_0^x A_n(x, \xi)(c_1 \cos \mu \xi + c_2 \sin \mu \xi) \, d\xi$$

where c_1 and c_2 denote two arbitrary constants.* On differentiating term by term (which is permissible) we obtain in addition to (70) the formula for the derivative

(70') $\qquad z'(x) = \mu(c_2 \cos \mu x - c_1 \sin \mu x) + \sum_{n=1}^{\infty} \mu^{-n} \int_0^x \frac{\partial A_n}{\partial x}(c_1 \cos \mu \xi + c_2 \sin \mu \xi) \, d\xi$

since $A_n(x, x) = 0$ for all n.

To deal with the boundary conditions we proceed as in § 22; as the fundamental interval is $(0, \pi)$ these conditions may be written in the two equivalent forms

(71) $\qquad hz(0) + h'z'(0) = 0, \qquad kz(\pi) + k'z'(\pi) = 0$

or

(71') $\qquad z(0) \cos \alpha - z'(0) \sin \alpha = 0, \qquad z(\pi) \cos \beta - z'(\pi) \sin \beta = 0$

where α and β are angles uniquely determined by the conditions

$$\tan \alpha = -h'/h \quad (0 \leq \alpha < \pi); \qquad \tan \beta = -k'/k \quad (0 < \beta \leq \pi)$$

Neglecting for the moment the boundary condition at the upper extreme $x = \pi$, since from (70) and (70') follow

$$z(0) = c_1, \qquad z'(0) = \mu c_2,$$

it is clear that the condition at the lower bound $x = 0$ will be satisfied if $c_1 = \sin \alpha$, $\mu c_2 = \cos \alpha$, i.e. if

(72) $\qquad z(x) = \sin \alpha \cos \mu x + \frac{1}{\mu} \cos \alpha \sin \mu x$

$$+ \sum_{n=1}^{\infty} \mu^{-n} \int_0^x A_n(x, \xi)(\sin \alpha \cos \mu \xi + \frac{1}{\mu} \cos \alpha \sin \mu \xi) \, d\xi$$

---

* The factor $1/\sqrt{\mu}$ is absorbed into the arbitrary constants.

This formula shows immediately that

(73) $$z(x) = \sin\alpha \cos\mu x + O(\mu^{-1})$$

a result of some considerable importance.* This result however can be much improved; for example we can obtain an asymptotic representation with residual term $O(\mu^{-2})$ instead of (as above) $O(\mu^{-1})$; this is done as follows.

From (72) we deduce

$$z(x) = \sin\alpha\cos\mu x + \frac{1}{\mu}\cos\alpha\sin\mu x + \frac{\sin\alpha}{\mu}\int_0^x \sin\mu(x-\xi)\cos\mu\xi\, Q(\xi)\, d\xi + O(\mu^{-2})$$

$$= \sin\alpha\cos\mu x + \frac{1}{\mu}\cos\alpha\sin\mu x + \frac{\sin\alpha}{2\mu}\int_0^x [\sin\mu x + \sin\mu(x-2\xi)]\, Q(\xi)\, d\xi + O(\mu^{-2})$$

$$= \sin\alpha\cos\mu x + \frac{1}{\mu}\left[\cos\alpha + \frac{\sin\alpha}{2}\int_0^x Q(\xi)\, d\xi\right]\sin\mu x$$

$$+ \frac{\sin\alpha\sin\mu x}{2\mu}\int_0^x Q(\xi)\cos 2\mu\xi\, d\xi - \frac{\sin\alpha\cos\mu x}{2\mu}\int_0^x Q(\xi)\sin 2\mu\xi\, d\xi + O(\mu^{-2})$$

But on the other hand, *if the function $Q(x)$ is of bounded variation,*** we have

(74) $$\int_0^x Q(\xi)\cos 2\mu\xi\, d\xi = O(\mu^{-1}), \quad \int_0^x Q(\xi)\sin 2\mu\xi\, d\xi = O(\mu^{-1})$$

and consequently the last two terms in the expression for $z(x)$ may be included in $O(\mu^{-2})$; thus

(75) $$z(x) = \sin\alpha\cos\mu x + \frac{1}{\mu}[\cos\alpha + T(x)\sin\alpha]\sin\mu x + O(\mu^{-2})$$

---

* This shows that for large values of μ the *sinusoidal* terms give a first approximation to the eigenfunctions.

** As any function of *bounded variation* may be represented as the sum of two bounded *monotonic* functions, and conversely, it is clearly sufficient to establish (74) under the hypothesis that $Q(x)$ be monotonic, in which case we can apply the Second Theorem of Mean Value—thus, if ϑ be any number between 0 and 1,

$$\int_0^x Q(\xi)\sin 2\mu\xi\, d\xi = Q(0)\int_0^{\vartheta x}\sin 2\mu\xi\, d\xi + Q(x)\int_{\vartheta x}^x \sin 2\mu\xi\, d\xi$$

$$= \frac{Q(0)}{2\mu}[1 - \cos 2\mu\vartheta x] + \frac{Q(x)}{2\mu}[\cos 2\mu\vartheta x - \cos 2\mu x] = O(\mu^{-1})$$

and a similar result holds for the cosine integral.

In order that a function be of bounded variation it is *sufficient* that it satisfies uniformly a Lipschitz condition.

where

(76) $$\tfrac{1}{2}\int_0^x Q(\xi)\,d\xi = T(x)$$

The formulæ (73) and (75), with which may be associated corresponding formulæ for the derivatives (deducible from (70′)),

(73′) $$z'(x) = -\mu\sin\alpha\sin\mu x + O(1)$$

(75′) $$z'(x) = -\mu\sin\alpha\sin\mu x + [\cos\alpha + T(x)\sin\alpha]\cos\mu x + O(\mu^{-1})$$

are generally more than sufficient for practical purposes provided that $\alpha \neq 0$, and they give *the first form of asymptotic expression for the eigenfunctions*.* If on the other hand $\alpha = 0$, it is convenient to replace $z(x)$ by $z(x)/\mu$ and (72) then gives, as far as the term $n = 1$,

$$z(x) = \sin\mu x + \frac{1}{\mu}\int_0^x \sin\mu(x-\xi)\sin\mu\xi\, Q(\xi)\,d\xi + O(\mu^{-2})$$

$$= \sin\mu x - \frac{\cos\mu x}{2\mu}\int_0^x Q(\xi)\,d\xi + \frac{1}{2\mu}\int_0^x \cos\mu(x-2\xi)Q(\xi)\,d\xi + O(\mu^{-2})$$

Hence, in view of (74) and (76), we obtain

(75″) $$z(x) = \sin\mu x - \frac{1}{\mu}T(x)\cos\mu x + O(\mu^{-2}) \qquad (\alpha = 0)$$

and

(75‴) $$z'(x) = \mu\cos\mu x + T(x)\sin\mu x + O(\mu^{-1}) \qquad (\alpha = 0)$$

It is obvious that if μ has any value (other than an eigenvalue—see the following section) the right-hand side of (75) multiplied by an arbitrary constant C yields an asymptotic expression as $\mu \to \infty$ for the general solution of (68) depending on the two arbitrary constants α and C.

34. Asymptotic expression for the eigenvalues

We shall now determine the 'eigenvalues' μ_n** of the problem making use of the boundary conditions at the upper limit $x = \pi$, i.e. we must solve for μ the equation

$$z(\pi)\cos\beta - z'(\pi)\sin\beta = 0$$

---

* The *first* form since the final form requires the (asymptotic) evaluation of the eigenvalues, i.e. of the values of μ.

** In fact the eigenvalues are the squares $\mu_1^2, \mu_2^2, \ldots$ and not $\mu_1, \mu_2, \ldots$ (cf. § 32); but for brevity we here call $\mu_1, \mu_2, \ldots$ the eigenvalues.

By making use of (75) and (75'), and dividing through by μ, this equation may be written in the form

$$\sin \alpha \sin \beta \sin \mu\pi + \{\sin \alpha \cos \beta - [\cos \alpha + T(\pi) \sin \alpha] \sin \beta\} \frac{\cos \mu\pi}{\mu} + O(\mu^{-2}) = 0$$

or, assuming that $\alpha > 0$, $\beta < \pi$, and dividing by $\sin \alpha \sin \beta$

(77) $$\sin \mu\pi + \frac{1}{\mu}[\cot \beta - \cot \alpha - T(\pi)]\cos \mu\pi + O(\mu^{-2}) = 0$$

It is helpful to apply to this equation a general theorem on the zeros of a function for which an asymptotic representation is known; this result stated in its simplest form is as follows:

If the function $f(x)$ may be represented, uniformly with respect to x, by the asymptotic formula

$$f(x) = g_0(x) + \lambda g_1(x) + O(\lambda^2)$$

where λ is any parameter dependent or independent of x (which parameter under certain conditions may affect also the functions g_0 and g_1), if the three derivatives $g_0'(x)$, $g_0''(x)$ and $g_1'(x)$ exist and the derivative $g_1'(x)$ is continuous, then with each simple root $x = x_0$ of the equation $g_0(x) = 0$ may be associated a root $x = x_0^*$ of the equation $f(x) = 0$ such that

(78) $$x_0^* = x_0 - \lambda \frac{g_1(x_0)}{g_0'(x_0)} + O(\lambda^2)$$

To establish this theorem, we begin by proving that if $(x_0 - \eta, x_0 + \eta)$ is a neighbourhood of x_0 in which $g_0'(x) \neq 0$—and we may assume $g_0'(x) > 0$ for this can always be arranged by changing the signs throughout the equation if necessary—then in any neighbourhood $(x_0 - \eta', x_0 + \eta')$ of x_0 where $\eta' \leq \eta$ there is one and only one root x_0^* of the equation $f(x) = 0$, provided that $|\lambda|$ is sufficiently small.

As $g_0(x)$ is an increasing function of x in the interval $(x_0 - \eta', x_0 + \eta')$ then $g_0(x_0 + \eta') > 0$ and $g_0(x_0 - \eta') < 0$. Also, $f(x) - g_0(x)$ tends to zero as $\lambda \to 0$; thus, for $|\lambda|$ less than some suitable λ_0, we can write

$$|f(x_0 + \eta') - g_0(x_0 + \eta')| < \tfrac{1}{2} g_0(x_0 + \eta')$$
$$|f(x_0 - \eta') - g_0(x_0 - \eta')| < \tfrac{1}{2} |g_0(x_0 - \eta')|$$

which implies $f(x_0 + \eta') > 0$, $f(x_0 - \eta') < 0$; therefore the function $f(x)$ vanishes in at least one point x_0^* in the interval $(x_0 - \eta', x_0 + \eta')$. Further, this zero of $f(x)$ is unique (provided that λ_0 is sufficiently small) since if there were two zeros x_0^* and x_0^{**}, where $x_0^* < x_0^{**}$ say, then

$$g_0(x_0^{**}) - g_0(x_0^*) = R(x_0^*, \lambda) - R(x_0^{**}, \lambda)$$

where $R(x, \lambda)$ denotes the difference $f(x) - g_0(x)$; but this is impossible as for sufficiently small values of $|\lambda|$ the left-hand side is a determinate positive quantity while the right-hand side tends to zero as $\lambda \to 0$.

To find the asymptotic expression for the difference $x_0^* - x_0$, we can start with the assertion that this difference is of order $O(\lambda)$, since for a suitable $\bar{x}_0$ in the interval (x_0, x) the equation $f(x) = 0$ may be written as

$$(x - x_0)g_0'(\bar{x}_0) + \lambda g_1(x) + O(\lambda^2) = 0$$

and consequently

$$x_0^* - x_0 = -\lambda \frac{g_1(x_0^*)}{g_0'(\bar{x}_0)} + O(\lambda^2)$$

To obtain the stated asymptotic expression for $x_0^* - x_0$ all that is now required is to note that since this difference (and therefore also the difference $\bar{x}_0 - x_0$) is $O(\lambda)$, then

$$g_1(x_0^*) = g_1(x_0) + O(\lambda), \qquad g_0'(\bar{x}_0) = g_0'(x_0) + O(\lambda)$$

hence

$$\frac{g_1(x_0^*)}{g_0'(\bar{x}_0)} = \frac{g_1(x_0)}{g_0'(x_0)} + O(\lambda)$$

and on substitution into the expression for x_0^* above we immediately obtain (78).

It should be noticed that when λ depends on x the value of λ to be substituted into the right-hand side of (78) ought to be that corresponding to $x = x_0^*$, but in fact it may be that corresponding to $x = x_0$ if $\log \lambda$ is lipschitz in x (for example, if $\log \lambda$ is differentiable with respect to x) since then

$$\log \lambda(x_0^*) = \log \lambda(x_0) + O(x_0^* - x_0) = \log \lambda(x_0) + O[\lambda(x_0^*)]$$

implying that

$$\lambda(x_0^*) = \lambda(x_0)\{1 + O[\lambda(x_0^*)]\}$$

and

$$\lambda(x_0^*)\frac{g_1(x_0)}{g_0'(x_0)} = \lambda(x_0)\frac{g_1(x_0)}{g_0'(x_0)} + O(\lambda^2)$$

We now apply this theorem to (77); we first observe that from the 'approximate' equation
$$\sin \mu \pi = 0$$
it must follow that
$$\mu_n^{(0)} = n \qquad (n = 1, 2, 3, \ldots)$$
if to the $(n+1)$th 'eigenvalue' μ_n* there corresponds an eigenfunction vanishing n times within the interval $(0, \pi)$ (cf. § 24)—in view of (73) and the inequality $0 < \alpha < \pi$. On the other hand, replacing λ by $1/\mu$ we have
$$g_0(\mu) = \sin \mu \pi, \qquad g_0'(\mu) = \pi \cos \mu \pi$$
$$g_1(\mu) = [\cot \beta - \cot \alpha - T(\pi)] \cos \mu \pi$$
whence, by (78), we may write

(79) $$\mu_n = n + \frac{1}{\pi n} [\cot \alpha - \cot \beta + T(\pi)] + O(n^{-2})$$

or

(79') $$\mu_n = n + \frac{1}{\pi n} \left[\frac{k}{k'} - \frac{h}{h'} + T(\pi) \right] + O(n^{-2})$$

These results are derived on the assumption that $\alpha > 0$, $\beta < \pi$.

If instead we have $\alpha > 0$, $\beta = \pi$, then (77) should be replaced by the equation $z(\pi) = 0$, which on division by $\sin \alpha$ takes the form
$$\cos \mu \pi + \frac{1}{\mu} [\cot \alpha + T(\pi)] \sin \mu \pi + O(\mu^{-2}) = 0$$
from which, as above, is derived the formula
$$\mu_n = n + \tfrac{1}{2} + \frac{1}{\pi(n+\tfrac{1}{2})} [\cot \alpha + T(\pi)] + O(n^{-2})$$

This may be written in the alternative forms

(80) $$\mu_n = n + \tfrac{1}{2} + \frac{1}{\pi n} [\cot \alpha + T(\pi)] + O(n^{-2}) \qquad (\beta = \pi)$$

or

(80') $$\mu_n = n + \tfrac{1}{2} + \frac{1}{\pi n} \left[-\frac{h}{h'} + T(\pi) \right] + O(n^{-2}) \qquad (\beta = \pi)$$

---

* Considering the first 'eigenvalue' to be that corresponding to $n = 0$ this is the $(n+1)$th 'eigenvalue'.

§ 34] *Asymptotic expression for the eigenvalues*

For the case in which $\alpha = 0$, $\beta < \pi$, in view of (75") and (75'''), the equation to be satisfied at the upper limit $x = \pi$ when divided by $-\mu \sin \beta \neq 0$ takes the form

$$\cos \mu\pi + \frac{1}{\mu}[-\cot \beta + T(\pi)]\sin \mu\pi + O(\mu^{-2}) = 0$$

Exactly as above we therefore find

(81) $$\mu_n = n + \tfrac{1}{2} + \frac{1}{\pi n}[-\cot \beta + T(\pi)] + O(n^{-2}) \qquad (\alpha = 0)$$

or

(81') $$\mu_n = n + \tfrac{1}{2} + \frac{1}{\pi n}\left[\frac{k}{k'} + T(\pi)\right] + O(n^{-2}) \qquad (\alpha = 0)$$

Lastly there remains to be considered the case for which $\alpha = 0$, $\beta = \pi$; here, from (75"),

$$\sin \mu\pi - \frac{1}{\mu} T(\pi) \cos \mu\pi + O(\mu^{-2}) = 0$$

and it is necessary to observe in solving the 'approximate' equation

$$\sin \mu\pi = 0$$

that in this case we must assume $\mu_n^{(0)} = n+1$ in place of $\mu_n^{(0)} = n$ if the corresponding eigenfunction is to vanish n times *within* the interval $(0, \pi)$. We therefore arrive at the formula

$$\mu_n = n + 1 + \frac{1}{\pi(n+1)} T(\pi) + O(n^{-2})$$

which may be written as

(82) $$\mu_n = n + 1 + \frac{1}{\pi n} T(\pi) + O(n^{-2}) \qquad (\alpha = 0, \beta = \pi)$$

As an aid to remembering these results it is helpful to notice that the formulæ relative to the three sub-cases $\beta = \pi$, $\alpha \neq 0$; $\alpha = 0$, $\beta \neq \pi$; $\alpha = 0$, $\beta = \pi$, may be considered as contained in those for the general case, i.e. those given by (79)-(79'), and derived from these by suppressing those terms $\cot \alpha$, $\cot \beta$ or h/h', k/k' which become infinite provided that $1/2$ is added to the first term n for each of the terms suppressed.

35. Second form of asymptotic expression for the eigenfunctions

We return briefly to the eigenfunctions of the Sturm-Liouville system to derive an asymptotic expression for the $(n+1)^{\text{th}}$ eigenfunction $z_n(x)$; this is easily obtained from (75) of § 33 by now writing in place of μ the asymptotic expression (79′) for μ_n. For brevity it is convenient to put

$$\frac{1}{\pi}\left[T(\pi) - \frac{h}{h'} + \frac{k}{k'}\right] = c$$

thus

$$\cos \mu_n x = \cos\left[nx + \frac{cx}{n} + O(n^{-2})\right] = \cos\left(nx + \frac{cx}{n}\right) + O(n^{-2})$$

$$= \cos nx - \frac{cx}{n}\sin nx - \frac{1}{2!}\left(\frac{cx}{n}\right)^2 \cos nx + \ldots + O(n^{-2})$$

$$= \cos nx - \frac{cx}{n}\sin nx + O(n^{-2})$$

and similarly

$$\sin \mu_n x = \sin nx + \frac{cx}{n}\cos nx + O(n^{-2})$$

But

$$\frac{1}{\mu_n} = \frac{1}{n} + O(n^{-2})$$

Thus from (75) we deduce

$$z_n(x) = \sin \alpha \left[\cos nx - \frac{cx}{n}\sin nx\right] + \frac{1}{n}[\cos \alpha + T(x)\sin \alpha]\sin nx + O(n^{-2})$$

$$= \sin \alpha \left\{\cos nx + \frac{1}{n}[T(x) + \cot \alpha - cx]\sin nx\right\} + O(n^{-2})$$

i.e.

(83) $$z_n(x) = \sin \alpha \left\{\cos nx + \frac{1}{n}\left[T(x) - \frac{h}{h'} - cx\right]\sin nx\right\} + O(n^{-2})$$

To see what modifications are required in these formulæ when instead of referring to a *particular* eigenfunction, (i.e. to the eigenfunction relative to a particular value of the factor of proportionality attached to it) we refer

§ 35] Second form of asymptotic expression for the eigenfunctions

to a *normalized* eigenfunction, we must begin by calculating the corresponding normalizing factor N (§ 26, p. 125); there putting $r(x) \equiv 1$, $a = 0$, $b = \pi$ we obtain

$$N^{-2} = \int_0^\pi z_n^2(x)\, dx$$

This integral may be calculated *asymptotically* by squaring and then integrating both sides of* (83); thus

$$N^{-2} = \sin^2\alpha \int_0^\pi \left\{\cos^2 nx + \frac{1}{n}\left[T(x) - \frac{h}{h'} - cx\right]\sin 2nx\right\} dx + O(n^{-2})$$

But as the function $T(x) - h/h' - cx$ is differentiable the integral of its product by $\sin 2nx$ is $O(n^{-1})$ (cf. p. 168), and therefore follows the simpler result

$$N^{-2} = \sin^2\alpha \int_0^\pi \cos^2 nx\, dx + O(n^{-2})$$

i.e.
$$N^{-2} = \frac{\pi}{2}\sin^2\alpha + O(n^{-2})$$

The asymptotic formula for the nth *normalized* eigenfunction $\phi_n(x)$ is therefore

(84) $$\phi_n(x) = \sqrt{\left(\frac{2}{\pi}\right)}\left\{\cos nx + \frac{1}{n}\left[T(x) - \frac{h}{h'} - cx\right]\sin nx\right\} + O(n^{-2})$$

provided that $\alpha \neq 0$, $\beta \neq \pi$, i.e. $h' \neq 0$, $k' \neq 0$. If, instead, we have $k' = 0$ but $h' \neq 0$, by exactly analogous calculations to those above we obtain the formula

(84') $$\phi_n(x) = \sqrt{\left(\frac{2}{\pi}\right)}\Big\{\cos(n+\tfrac{1}{2})x$$
$$+ \frac{1}{n}\left[T(x) - \frac{h}{h'} - c'x\right]\sin(n+\tfrac{1}{2})x\Big\} + O(n^{-2}) \quad (h' \neq 0,\; k' = 0)$$

* If
$$F(x) = f(x) + O(n^{-h})$$
i.e.
$$F(x) = f(x) + \frac{A(x)}{n^h}$$
where $A(x)$ is a bounded function, then
$$\int_a^b F(x)\, dx = \int_a^b f(x)\, dx + O(n^{-h})$$
since
$$\int_a^b [F(x) - f(x)]\, dx = \frac{1}{n^h}\int_a^b A(x)\, dx = O(n^{-h})$$

where
$$\frac{1}{\pi}\left[T(\pi)-\frac{h}{h'}\right] = c'$$

On the other hand the results are not exactly similar in the case $h' = 0$ (whether or not $k' = 0$) since then the first term of (75) vanishes and we require to use (75''); this together with (81') gives, for $k' \neq 0$,

(83'') $$z_n(x) = \sin(n+\tfrac{1}{2})x - \frac{1}{n}[T(x) - c''x]\cos(n+\tfrac{1}{2})x + O(n^{-2})$$

where
$$c'' = \frac{1}{\pi}\left[T(\pi) + \frac{k}{k'}\right]$$

In the subcase $h' = k' = 0$, we find from (82),

(83''') $$z_n(x) = \sin(n+1)x - \frac{1}{n}[T(x) - c'''x]\cos(n+1)x + O(n^{-2})$$

where
$$c''' = \frac{1}{\pi}T(\pi)$$

In both these subcases the normalization factor is given by
$$N^{-2} = \frac{\pi}{2} + O(n^{-2})$$

so that

(84'') $$\phi_n(x) = \sqrt{\left(\frac{2}{\pi}\right)}\left\{\sin(n+\tfrac{1}{2})x - \frac{1}{n}[T(x) - c''x]\cos(n+\tfrac{1}{2})x\right\} + O(n^{-2})$$
$$(h' = 0,\ k' \neq 0)$$

and

(84''') $$\phi_n(x) = \sqrt{\left(\frac{2}{\pi}\right)}\left\{\sin(n+1)x - \frac{1}{n}[T(x) - c'''x]\cos(n+1)x\right\} + O(n^{-2})$$
$$(h' = k' = 0)$$

By using (73') and (75') we can derive a similar set of asymptotic formulæ for the first derivatives of the eigenfunctions; these are

§35] Second form of asymptotic expression for the eigenfunctions

$$
(85) \begin{cases} \phi'_n(x) = \sqrt{\left(\dfrac{2}{\pi}\right)}\left\{-n\sin nx + \left[T(x) - \dfrac{h}{h'} - cx\right]\cos nx\right\} + O(n^{-1}) \\ \hspace{10cm} (h' \neq 0,\ k' \neq 0) \\[4pt] \phi'_n(x) = \sqrt{\left(\dfrac{2}{\pi}\right)}\left\{-(n+\tfrac{1}{2})\sin(n+\tfrac{1}{2})x + \left[T(x) - \dfrac{h}{h'} - c'x\right]\cos(n+\tfrac{1}{2})x\right\} \\ \hspace{6cm} + O(n^{-1}) \quad (h' \neq 0,\ k' = 0) \\[4pt] \phi'_n(x) = \sqrt{\left(\dfrac{2}{\pi}\right)}\{(n+\tfrac{1}{2})\cos(n+\tfrac{1}{2})x + [T(x) - c''x]\sin(n+\tfrac{1}{2})x\} \\ \hspace{6cm} + O(n^{-1}) \quad (h' = 0,\ k' \neq 0) \\[4pt] \phi'_n(x) = \sqrt{\left(\dfrac{2}{\pi}\right)}\{(n+1)\cos(n+1)x + [T(x) - c'''x]\sin(n+1)x\} \\ \hspace{6cm} + O(n^{-1}) \quad (h' = k' = 0) \end{cases}
$$

It is of interest to note that, as for (75′) and (75‴), these new results are equivalent to those which could have been obtained, without rigour, by differentiating the asymptotic expressions for the eigenfunctions and changing the residual term $O(n^{-2})$ into $O(n^{-1})$. The above calculations therefore justify these differentiations which might *a priori* be regarded with suspicion.

36. Equations with transition points

We shall now deal briefly with the asymptotic behaviour as $\mu \to \infty$ of the integrals of the canonical equation (68′) in the vicinity of the point $x = 0$; this point is not a singular point of (68′) in the ordinary sense of the word but is a 'privileged' point as regards the asymptotic representation of its integrals; indeed the sign of

$$P(x) = \mu^2 x - Q(x)$$

which determines (see § 19) whether the integrals are oscillatory or not, for large values of μ depends essentially on the term $\mu^2 x$ which changes sign as x passes through the origin. A point of this nature will be called a *transition point*.*

The more direct method for the study of (68′) is that of Fubini (§ 29); we use the change of variable

(86) $$x = 3^{-1/3}\mu^{-2/3} t$$

---

* With reference to the equation (64) and under the condition that $(px) > 0$, the transition points are the zeros of odd order of the function $r(x)$. On this subject see a paper by A. ERDÉLYI: *Proc. Inter. Congress Math., Amsterdam*, 3, 92–101 (1956).

which transforms equation (68′) into

(87) $$\frac{d^2z}{dt^2}+\tfrac{1}{3}tz = \mu^{-4/3}Q^*(t)z$$

where

(88) $$Q^*(t) = 3^{-2/3}Q(3^{-1/3}\mu^{-2/3}t)$$

in which, assuming that the function $Q(x)$ is bounded (at least within a convenient neighbourhood of the origin) and that t is bounded, the coefficient on the right-hand side is $O(\mu^{-4/3})$. Further, the 'approximate' equation

(89) $$\frac{d^2z}{dt^2}+\tfrac{1}{3}tz = 0$$

is that satisfied by the Airy function

$$A_1(t) = \frac{\pi}{3}\sqrt{\left(\frac{t}{3}\right)}\left\{J_{-1/3}\left[2\left(\frac{t}{3}\right)^{3/2}\right]+J_{1/3}\left[2\left(\frac{t}{3}\right)^{3/2}\right]\right\}$$

and *its associated function**

$$A_2(t) = \frac{\pi}{3}\sqrt{t}\left\{J_{-1/3}\left[2\left(\frac{t}{3}\right)^{3/2}\right]-J_{1/3}\left[2\left(\frac{t}{3}\right)^{3/2}\right]\right\}$$

which is linearly independent of $A_1(t)$ (for the Wronskian formed by these two functions equals $-\pi/3$).

Employing (35) we can now assert that *the canonical equation* (68′) *in which the function $Q(x)$ is continuous in the neighbourhood of the origin possesses two linearly independent integrals Z_1 and Z_2 having the asymptotic representations*

$$Z_h = A_h(t)+O(\mu^{-4/3}) \qquad (h = 1, 2)$$

i.e.

(90) $$Z_h = A_h(3^{1/3}\mu^{2/3}x)+O(\mu^{-4/3})$$

This result justifies the name *transition point* given to the point $x = 0$, since the argument of A_h tends to $+\infty$ or to $-\infty$ as $\mu\to\infty$ according as $x>0$ or $x<0$, and the asymptotic behaviour of the functions $A_h(t)$ as $t\to+\infty$ is completely different from that occurring when $t\to-\infty$.

---

* This function was used by the author in his paper: 'Sul comportamento asintotico dell'$n$esimo polinomio di Laguerre nell'intorno dell'ascissa $4n$', *Comm. Math. Helvetici*, 22, 150–167 (1949).

In the first case, employing the classical asymptotic formulæ for the Bessel functions which are to be found in the last section (§ 52) of this book, we obtain the result

(91) $$A_h(t) = \sqrt{\pi}(3t)^{-1/4} \cos\left[2\left(\frac{t}{3}\right)^{3/2} \mp \frac{\pi}{4}\right] + O(t^{-7/4})$$

where the $-$ sign refers to the case $h = 1$ and the $+$ sign to the case $h = 2$. This formula clearly shows the (damped) oscillatory character of the two functions as $t \to +\infty$. On the other hand, for $t \to -\infty$ the function $A_1(t)$ tends (exponentially) to *zero* and the function $A_2(t)$ tends (also exponentially) to *infinity* as is shown by the formulæ

(92) $$\begin{cases} A_1(-t) = \tfrac{1}{2}\sqrt{\pi}(3t)^{-1/4} \exp\left[-2\left(\frac{t}{3}\right)^{3/2}\right] \cdot [1 + O(t^{-3/2})] \\ A_2(-t) = \sqrt{\pi}(3t)^{-1/4} \exp\left[2\left(\frac{t}{3}\right)^{3/2}\right] \cdot [1 + O(t^{-3/2})] \end{cases}$$

derived from the same asymptotic representations of the Bessel functions but most easily obtained from the expressions

(93) $$\begin{cases} A_1(-t) = \tfrac{1}{3}\sqrt{t}\, K_{1/3}\left[2\left(\frac{t}{3}\right)^{3/2}\right] \\ A_2(-t) = \frac{\pi}{3}\sqrt{t}\left\{I_{-1/3}\left[2\left(\frac{t}{3}\right)^{3/2}\right] + I_{1/3}\left[2\left(\frac{t}{3}\right)^{3/2}\right]\right\} \end{cases}$$

where K_ν and I_ν denote *modified Bessel functions.**

When the approximations stated in (90) are insufficient we require to calculate successive terms of the asymptotic expansion of Z_h by the method outlined in § 29. For example, to obtain an asymptotic formula containing two terms, we use (35') which leads here to the equation

$$Z_h = A_h(t) + \int_0^t L(t, \tau) \mathcal{G}_h(\tau)\, d\tau + O(\mu^{-8/3})$$

where

$$\mathcal{G}_h(t) = \frac{Q^*(t) A_h(t)}{-\pi/3} \mu^{-4/3}$$

and

$$L(t, \tau) = A_1(\tau) A_2(t) - A_1(t) A_2(\tau)$$

---

* See, for example, W. MAGNUS-F. OBERHETTINGER (69), p. 28.

If now we put

(94) $$\int_0^t Q^*(\tau) A_h(\tau) A_k(\tau) \, d\tau = B_{h,k}(t)$$

the formula for Z_h may be written as

(95) $$Z_h = A_h(t) + \frac{3}{\pi} \mu^{-4/3} [A_1(t) B_{h,2}(t) - A_2(t) B_{h,1}(t)] + O(\mu^{-8/3})$$

We omit the transformation to the original variable x by means of (86).

37. The Laguerre differential equation and polynomials

As has been earlier pointed out in this chapter in some applications of the theory of differential equations, as for example in some applications in atomic physics, the asymptotic behaviour of the solution of the equation under consideration is specified by the data of the problem. For example, if the fundamental interval be the interval $(0, +\infty)$ it may happen that the solution required, with prescribed initial value $y(0)$ at the origin, is subjected to the single additional condition of being bounded as $x \to +\infty$, or of not increasing more rapidly than a certain power of x, etc. In other cases it may happen that the fundamental interval is the finite interval (a, b) and the desired solution is to remain bounded at the extremes of this interval although these points may be singular points of the differential equation, etc.

In view of the multiplicity of possible cases and the consequent difficulty of incorporating them into a general theory we shall limit ourselves here to developing some particular examples which, in addition to being of interest in themselves, are illustrative of the main methods so far developed for tackling problems of these types.

It is worth while observing here that the general correspondence existing between proper Sturm-Liouville systems and those with additional asymptotic conditions does not debar the existence of several differences; in particular it does not exclude the existence of a *continuous spectrum* of eigenvalues for the latter systems, i.e. a set of eigenvalues covering entirely one or more segments of the λ-axis.

This happens, for example, in the elementary case of the equation with constant coefficients

$$y'' + \lambda y = 0$$

considered in the interval $(0, +\infty)$ together with the conditions $y(0) = 0$ and *that $y(x)$ remains bounded as $x \to +\infty$*.

The general integral of the equation is

$$y = C_1 \sin \sqrt{(\lambda)}\, x + C_2 \cos \sqrt{(\lambda)}\, x, \qquad y = C_1 x + C_2$$

$$y = C_1 \sinh \sqrt{(-\lambda)}\, x + C_2 \cosh \sqrt{(-\lambda)}\, x$$

according as $\lambda > 0$, $\lambda = 0$, or $\lambda < 0$; in the first case to satisfy the boundary conditions it is sufficient to put $C_2 = 0$, while in the second and third cases the first boundary condition is similarly satisfied by putting $C_2 = 0$, but the other condition cannot be satisfied other than by putting $C_1 = 0$ which leads to the trivial solution $y \equiv 0$. We therefore conclude that *all the positive values of λ are eigenvalues of the problem under consideration and these are the only eigenvalues:* the eigenvalues therefore form a *continuous spectrum* in the sense indicated above.

We now consider the most interesting, although not the simplest, example of *Laguerre's equation*,* viz.

(96) $$xy'' + (\alpha + 1 - x)y' + \lambda y = 0$$

where α is a real constant which we suppose *greater than* -1 and λ a real parameter, *a priori* unknown. The equation is considered in the interval $(0, +\infty)$ and we look for solutions y satisfying the following asymptotic conditions at the extremes:

(1) *regular in the vicinity of $x = 0$*, i.e. y is representable by a power series in x within a suitable interval to the right of the origin.

(2) *increasing algebraically as $x \to +\infty$*, i.e. there must exist a positive integer N so large that to the right of a certain x_0

(97) $$|y(x)| < A x^N$$

where A is a suitable positive constant.

In view of the first condition we write

(98) $$y = c_0 + c_1 x + c_2 x^2 + c_3 x^3 + \dots$$

and substitute into the given equation to determine the coefficients c_ν. This gives

$$\lambda c_0 + (\alpha + 1)c_1 + \sum_{\nu=1}^{\infty} [(\lambda - \nu)c_\nu + (\nu + 1)(\nu + \alpha + 1)c_{\nu+1}]x^\nu = 0$$

---

* This is a 'shorthand' notation for *the equation satisfied by the Laguerre polynomials*. As will be seen in the next chapter (§ 52), equation (96) coincides with the differential equation satisfied by the *confluent hypergeometric functions*.

so that the coefficients must satisfy the recurrence relations

(99) $$c_{\nu+1} = -\frac{\lambda-\nu}{(\nu+1)(\nu+\alpha+1)} c_\nu \qquad (\nu = 0, 1, 2, \ldots)$$

whence

(100) $$c_\nu = (-1)^\nu \frac{\lambda(\lambda-1)\ldots(\lambda-\nu+1)}{\nu!(\alpha+1)(\alpha+2)\ldots(\alpha+\nu)} c_0$$

These results show that there are ∞^1 possible solutions of the Laguerre equation satisfying the first of the imposed conditions, for from (99) it follows that

$$\lim_{\nu\to\infty} \frac{c_\nu}{c_{\nu+1}} = \lim_{\nu\to\infty} \frac{(\nu+1)(\nu+\alpha+1)}{\nu-\lambda} = \infty$$

and therefore the series (98) has *infinite radius of convergence*.

However this solution does not in general satisfy the second condition—for *excluding the case in which λ is a non-negative integer and therefore for which the series (98) is finite and reduces to a polynomial*, it is impossible to find values of N and A for which the condition (97) is satisfied.

Since $(\nu+1)(\nu+\alpha+1)>0$ and since, whatever be the value (real) of λ, from a certain ν_0 onwards $\nu-\lambda>0$, (99) implies that for $\nu>\nu_0$, c_ν and $c_{\nu+1}$ have the same sign; hence, we may suppose that for $\nu>\nu_0$, $c_\nu>0$ (by changing the sign of c_0 if necessary). Consequently, dividing the series (98) into two parts and writing

$$P(x) = \sum_{\nu=0}^{\nu_0} c_\nu x^\nu, \qquad F(x) = \sum_{\nu=\nu_0+1}^{\infty} c_\nu x^\nu$$

we may consider the solution $y(x)$ as the sum of a certain polynomial $P(x)$ of degree ν_0 and a series representing an integral function $F(x)$ *in which all the coefficients are positive*. But such a series evidently cannot satisfy a boundary condition of the type (97) since its sum is greater than any one of its terms and the sum contains terms of degree as large as we please. It is therefore impossible that the solution $y(x)$ should satisfy (97), for otherwise the function $F(x) = y(x) - P(x)$ must satisfy a similar boundary condition, in fact one in which the index of the power of x on the right-hand side is the greater of the two numbers N and ν_0.

This discussion has assumed that λ is not a non-negative integer. If instead we have $\lambda=n$, $(n=0, 1, 2, 3, \ldots)$, then $c_{n+1}=c_{n+2}=\ldots=0$ and the series (98) reduces to a polynomial of degree n which obviously satisfies the boundary condition (97) with $N = n$.

Hence we conclude that *the problem under consideration has a solution if and only if the parameter λ has one of the values $0, 1, 2, 3, \ldots$ which will also be called eigenvalues.*

The *eigenfunction* corresponding to the $(n+1)^{\text{th}}$ eigenvalue $\lambda_n = n$ is a polynomial of degree n in x which we shall completely determine (as yet it has been determined only as far as a constant factor) by writing

(101) $$c_n = \frac{(-1)^n}{n!}$$

This* implies

$$c_0 = \frac{(\alpha+1)(\alpha+2)\ldots(\alpha+n)}{n!}$$

and, by (100),

$$c_\nu = (-1)^\nu \frac{(\alpha+n)(\alpha+n-1)\ldots(\alpha+\nu+1)}{(n-\nu)!\,\nu!} = \frac{(-1)^\nu}{\nu!}\binom{n+\alpha}{n-\nu}$$

$$(\nu = 0, 1, 2, \ldots, n)$$

The polynomial obtained by inserting these coefficients into $P(x)$ is called the *Laguerre polynomial* (of degree n) and is denoted by the symbol $L_n^{(\alpha)}(x)$; its explicit expression is

(102) $$L_n^{(\alpha)}(x) = \sum_{\nu=0}^{n} (-1)^\nu \binom{n+\alpha}{n-\nu} \frac{x^\nu}{\nu!}$$

In particular, when $\alpha = 0$ (this gives the *classic* Laguerre polynomials) we have

(102') $$L_n(x) = \sum_{\nu=0}^{n} (-1)^\nu \binom{n}{\nu} \frac{x^\nu}{\nu!}$$

The first few of these last polynomials are

$$L_0(x) = 1, \quad L_1(x) = 1-x$$
$$L_2(x) = 1-2x+\tfrac{1}{2}x^2, \quad L_3(x) = 1-3x+\tfrac{3}{2}x^2-\tfrac{1}{6}x^3$$

The Laguerre polynomials are of considerable importance in various questions in mathematics, both pure and applied**; they possess well-known

---

* Many authors define c_n as $(-1)^n$ and the resulting polynomials therefore differ by the factor $n!$ from those considered here.

** See, for example, E. PERSICO: *Fondamenti della meccanica atomica*, pp. 230 onwards (Bologna, Zanichelli, 1940).

Included in the Laguerre polynomials as particular cases are the *Hermite polynomials* $H_\nu(x)$; these are derived from the Laguerre polynomials according as the integer ν is even or odd by the relations

$$H_{2n}(x) = (-2)^n n!\, L_n^{(-1/2)}\!\left(\frac{x^2}{2}\right); \quad H_{2n+1}(x) = (-2)^n n!\, L_n^{(1/2)}\!\left(\frac{x^2}{2}\right)$$

For discussion of these polynomials (which many writers treat independently of the Laguerre polynomials) see, for example, G. VITALI-G. SANSONE (84), or G. SZEGÖ (78).

properties most of which arise from the fact that they are the eigenfunctions of (96) where the boundary conditions are those stated above.*

For example, they satisfy an *orthogonality relation* which is easily obtained** by writing (96) (in which λ has been replaced by n) in the self-adjoint form derived by multiplying throughout by $x^\alpha e^{-x}$; thus

$$\frac{d}{dx}(x^{\alpha+1}e^{-x}y') + nx^\alpha e^{-x}y = 0$$

from which follows the identity

$$L_m^{(\alpha)}(x)\frac{d}{dx}[x^{\alpha+1}e^{-x}L_n^{(\alpha)\prime}(x)] - L_n^{(\alpha)}(x)\frac{d}{dx}[x^{\alpha+1}e^{-x}L_m^{(\alpha)\prime}(x)]$$
$$= (m-n)x^\alpha e^{-x}L_m^{(\alpha)}(x)L_n^{(\alpha)}(x)$$

i.e.

$$\frac{d}{dx}\{x^{\alpha+1}e^{-x}[L_m^{(\alpha)}(x)L_n^{(\alpha)\prime}(x) - L_m^{(\alpha)\prime}(x)L_n^{(\alpha)}(x)]\} = (m-n)x^\alpha e^{-x}L_m^{(\alpha)}(x)L_n^{(\alpha)}(x)$$

We now integrate both sides of this identity in a finite interval (a, b) where $0 < a < b$, then make a tend to zero and b tend to $+\infty$; in view of the first boundary condition on the Laguerre polynomials and since $\alpha + 1 > 0$, it follows that

$$\lim_{a \to 0}\{a^{\alpha+1}e^{-a}[L_m^{(\alpha)}(a)L_n^{(\alpha)\prime}(a) - L_m^{(\alpha)\prime}(a)L_n^{(\alpha)}(a)]\} = 0$$

Also, in view of the second condition on the Laguerre polynomials, the limit as $b \to +\infty$ of the analogous expression in which b is written in place of a must be zero, as the exponential factor is the dominating factor since the other factors are polynomials. It therefore follows that for $m \neq n$

(103) $$\int_0^\infty x^\alpha e^{-x} L_m^{(\alpha)}(x) L_n^{(\alpha)}(x)\, dx = 0$$

which is the orthogonality relation.

It can be easily deduced from the orthogonality relation (103) that *all the zeros of $L_n^{(\alpha)}(x)$ are real, simple and contained within the fundamental interval* $(0, +\infty)$.

---

* These conditions may be weakened. For a discussion of this, see the two papers of M. PICONE and G. SANSONE both entitled, 'I polinomi di Hermite e di Laguerre come autosoluzioni', *Boll. Unione Mat. Ital.*, respectively (1) **16**, 205–18 (1937), and (2) **2**, 193–200 (1940).

** This relation cannot be immediately deduced from the results of p. 124 since here the fundamental interval is infinite and the boundary conditions are of a kind different from those used in Chapter III.

In fact there is no greater complication in proving the more general result, viz.

If in a certain interval (a, b) (finite or infinite) the polynomials $P_0(x)$, $P_1(x)$, $P_2(x)$, ..., of degrees equal to their suffices satisfy the orthogonality relations

$$\int_a^b r(x) P_m(x) P_n(x) \, dx = 0 \qquad (m \neq n)$$

where $r(x)$ denotes a function of constant sign in (a, b), then each of the functions P_i has all its zeros real, simple and contained between a and b.

If $x_1, x_2, \ldots, x_m (0 \leq m \leq n)$ are the real (distinct) zeros of $P_n(x)$ contained between a and b which are either simple or multiple of *odd* order, so that on passing through these zeros $P_n(x)$ changes sign, then $P_n(x)$ changes sign as x varies from a to b if and only if this is true for the polynomial of degree m

$$Q_m(x) = (x-x_1)(x-x_2) \ldots (x-x_m)$$

Hence the product

$$r(x) P_n(x) Q_m(x)$$

is of constant sign in the entire interval (a, b), which implies that

(104) $$I = \int_a^b r(x) P_n(x) Q_m(x) \, dx \neq 0$$

But since $P_0(x), P_1(x), \ldots$, are polynomials of degrees equal to their suffices, any polynomial Q_m of degree m may be represented as a linear combination of $P_0(x), P_1(x), \ldots, P_m(x)$,* i.e. there exist suitable constants $a_0, a_1, \ldots, a_m$ such that

$$Q_m(x) \equiv a_0 P_0(x) + a_1 P_1(x) + \ldots + a_m P_m(x)$$

Hence

$$I = a_0 \int_a^b r(x) P_n(x) P_0(x) \, dx + a_1 \int_a^b r(x) P_n(x) P_1(x) \, dx + \ldots$$
$$+ a_m \int_a^b r(x) P_n(x) P_m(x) \, dx$$

If $m < n$ this is in contradiction to (104), since by the orthogonality relation all the integrals on the right-hand side are zero and therefore $I = 0$. Consequently $m = n$ and the theorem follows, since if a polynomial of degree n has n distinct zeros these must all be simple zeros.

* If α and β are the coefficients of x^m in $P_m(x)$ and $Q_m(x)$ respectively the difference $Q_m(x) - (\beta/\alpha)P_m(x)$ is a polynomial of degree $m - 1$; a similar argument can now be applied to this polynomial, and so on.

Among the numerous formulæ for Laguerre polynomials we point out only the following,

$$L_n^{(\alpha)}(x) = \frac{e^x x^{-\alpha}}{n!} \frac{d^n}{dx^n}(e^{-x} x^{n+\alpha})$$

$$(n+1)L_{n+1}^{(\alpha)}(x) - (2n+\alpha+1-x)L_n^{(\alpha)}(x) + (n+\alpha)L_{n-1}^{(\alpha)}(x) = 0$$

$$x\frac{d}{dx}L_n^{(\alpha)}(x) = nL_n^{(\alpha)}(x) - (n+\alpha)L_{n-1}^{(\alpha)}(x)$$

for whose proofs we refer the reader to works more directly concerned with these functions.*

38. Asymptotic behaviour of the Laguerre polynomials

In the majority of the problems in which the Laguerre polynomials arise the main interest lies in determining their asymptotic behaviour for large values of n; this is not easy as it is necessary to distinguish within the interval of orthogonality $(0, \infty)$ four distinct regions, these being (1) the neighbourhood immediately to the right of the origin; (2) the *open* interval (i.e. the extremes excluded) from the origin to the transition point $x = 4n_1 = 4n + 2(\alpha+1)$; (3) the neighbourhood about $x = 4n_1$; (4) the part of the x-axis to the right of the point $x = 4n_1$.** It is of considerable importance that the general method outlined in § 29 provides a simple asymptotic representation derived from (96) of $L_n^{(\alpha)}(x)$ valid either in the first or third of the four regions above, which within the required limits gives fairly satisfactory results.

We begin with the vicinity of the origin; the transformation

$$y(x) = e^{hx} z(x)$$

transforms equation (96) into

$$xz'' + [\alpha+1+(2h-1)x]z' + [(\alpha+1)h + n - h(1-h)x]z = 0$$

which takes a very simple form in the particular case $h = 1/2$. Putting $h = 1/2$ and writing

(105) $$n + \frac{\alpha+1}{2} = n_1$$

---

* See G. Szegö (78); also G. Vitali-G. Sansone (84), and F. Tricomi (81).
** For an exhaustive study of this question, see F. Tricomi: 'Sul comportamento asintotico dei polinomi di Laguerre', *Ann. di Mat.*, (4) **28**, 263–89 (1949).

equation (96) becomes

(106) $$xz'' + (\alpha+1)z' + (n_1 - \tfrac{1}{4}x)z = 0$$

On making the further transformation $x = t/n_1$ and carrying the last term to the right-hand side, (96) becomes finally

(107) $$t\frac{d^2z}{dt^2} + (1+\alpha)\frac{dz}{dt} + z = \frac{t}{4n_1^2}z$$

This equation which has the form of (10) of § 29 is easily dealt with by using the method shown in § 29, partly because the coefficient of the term on the right-hand side is $O(n^{-2})$ as $n \to \infty$, and partly because the equation obtained by equating to zero the left-hand side, viz.

(107') $$t\frac{d^2z}{dt^2} + (1+\alpha)\frac{dz}{dt} + z = 0$$

is a well-known transformation of Bessel's equation ((32) of Chapter III), as is easily verified by making the changes of variables (in (32) of Chapter III, where $\nu = \alpha$),

$$x = 2\sqrt{t}, \qquad y = t^{\alpha/2}z$$

It follows that an integral of (107') is the function

$$F_1(t) = E_\alpha(t) \equiv t^{-\alpha/2} J_\alpha(2\sqrt{t})$$

represented by the series (cf. § 30) convergent for all finite t,

$$E_\alpha(t) = \frac{1}{\Gamma(\alpha+1)} - \frac{1}{\Gamma(\alpha+2)}\frac{t}{1!} + \frac{1}{\Gamma(\alpha+3)}\frac{t^2}{2!} - \cdots$$

while a second linearly independent integral is

$$F_2(t) = t^{-\alpha/2} N_\alpha(2\sqrt{t})$$

where N_α denotes the Bessel function of the second kind referred to in § 30.*

We can therefore assert, by (35) of § 29, that with every solution of Laguerre's equation—and therefore in particular with $L_n^{(\alpha)}(x)$—may be associated two constants γ_1 and γ_2 such that

(108) $$e^{-x/2} L_n^{(\alpha)}(x) = \gamma_1 [E_\alpha(n_1 x) + O(n^{-2})]$$
$$+ \gamma_2 \{(n_1 x)^{-\alpha/2} N_\alpha(2\sqrt{[n_1 x]}) + O(n^{-2})\}$$

---

* When α is not integral we may assume $F_2(t) = t^{-\alpha} E_{-\alpha}(t)$, since equation (107') is unaltered, apart from the change of α into $-\alpha$, by the substitution $z = t^{-\alpha}u$.

The rigorous determination of the two constants γ_1 and γ_2 is not easy. However it is fairly obvious* that (neglecting terms of order n^{-2}) at least when $\alpha > 0$ we must have γ_2 zero, for otherwise the explicit terms on the right-hand side of (108) would tend to infinity as $x \to 0$ while the left-hand side tends to the finite quantity

$$(109) \qquad L_n^{(\alpha)}(0) = \binom{n+\alpha}{n}$$

We shall assume that $\gamma_2 = 0$ also when $\alpha > -1$; by (109) this implies

$$\binom{n+\alpha}{n} = \gamma_1 E_\alpha(0) = \frac{\gamma_1}{\Gamma(\alpha+1)}$$

as far as terms of order n^{-2}, from which follows

$$\gamma_1 = \Gamma(\alpha+1)\binom{n+\alpha}{n} = \Gamma(\alpha+1)\frac{(\alpha+1)(\alpha+2)\ldots(\alpha+n)}{n!} = \frac{\Gamma(\alpha+n+1)}{n!}$$

We therefore reach the fairly simple asymptotic representation

$$(110) \qquad L_n^{(\alpha)}(x) = \frac{\Gamma(\alpha+n+1)}{n!} e^{x/2}[E_\alpha(n_1 x) + O(n^{-2})]$$

$$= \frac{\Gamma(\alpha+n+1)}{n!} e^{x/2}\{(n_1 x)^{-\alpha/2} J_\alpha(2\sqrt{[n_1 x_1]}) + O(n^{-2})\}$$

which may be further simplified by using the well-known property of the Gamma function**

$$\frac{\Gamma(\alpha+n+1)}{n!} = \frac{\Gamma\left(n_1 + \frac{1+\alpha}{2}\right)}{\Gamma\left(n_1 + \frac{1-\alpha}{2}\right)} = n_1^\alpha [1 + O(n_1^{-2})]$$

Thus

$$(111) \qquad L_n^{(\alpha)}(x) = e^{x/2} n_1^\alpha [E_\alpha(n_1 x) + O(n^{-2})]$$

provided that α and $n_1 x$ remain bounded as n tends to ∞.

---

*It is not in general true that a function and its asymptotic representation behave similarly (for example, both remain finite) in a passage to the limit, if this limit be different from that to which the asymptotic representation refers.

**Cf. F. TRICOMI: *Rend. Semin. Mat. Torino*, 9, 343–50 (1949–50), or F. TRICOMI-A. ERDÉLYI: *Pacific J. of Math.*, 1, 133–42 (1951).

In the following table several exact values of

$$e^{-x/2} L_{10}(x)$$

(the case $\alpha = 0$) are compared with the approximate values given by (111) when the residual term is neglected.

$x =$	0	0·1	0·2	0·3	0·4	0·5
Exact value	1·0000	0·1970	−0·2232	−0·3865	−0·3900	−0·3033
Approximate value	1·0000	0·1955	−0·2237	−0·3864	−0·3888	−0·3006

Although 10 is not a large value of n agreement between these two sets of values of $e^{-x/2} L_{10}(x)$ is good and in graphical representation to the usual scale it is practically impossible to distinguish the exact curve from the approximate one.

We next consider the asymptotic behaviour of $L_n^{(\alpha)}(x)$ in the third of the regions, i.e. in the vicinity of the transition point $x = 4n_1$.*

For this we make the change of variable

(112) $$x = 4n_1 - \left(\frac{16}{3} n_1\right)^{1/3} t$$

Equation (106) then becomes

(113) $$\frac{d^2 z}{dt^2} + \frac{1}{3} tz = \left(\frac{4}{3}\right)^{1/3} \left[t \frac{d^2 z}{dt^2} + (\alpha+1) \frac{dz}{dt}\right] (4n_1)^{-2/3}$$

and the corresponding 'approximate' equation is equation (89) of § 36.

As the coefficients on the right-hand side are $O(n^{-2/3})$ as $n \to \infty$, we need only insert the solutions $A_1(t)$ and $A_2(t)$ of (89) into formula (35) of § 29 to obtain the asymptotic representation

$$e^{-x/2} L_n^{(\alpha)}(x) = \gamma_1 [A_1(t) + O(n^{-2/3})] + \gamma_2 [A_2(t) + O(n^{-2/3})]$$

As before, the major difficulty is the determination of the two constants γ_1 and γ_2; for this we refer the reader to a recent work of the author** in which it is proved that

$$\gamma_1 = (-1)^n \frac{2^{-\alpha}}{\pi} \left(\frac{3}{2}\right)^{1/3} n_1^{-1/3} + O(n^{-1}), \qquad \gamma_2 = O(n^{-1})$$

---

* This abscissa is an end-point of the *oscillatory region* for $L_n^{(\alpha)}(x)$, i.e. the region in which its zeros can occur. It can be shown that all the zeros of $L_n^{(\alpha)}(x)$ lie to the left of the abscissa $4n_1 - 3/2$.

** F. TRICOMI: 'Sul comportamento asintotico dell'nesimo polinomio di Laguerre nell'intorno dell'ascissa 4n', *Comm. Math. Helvetici*, **22**, 263–89 (1949).

Thus

(114) $$e^{-x/2} L_n^{(\alpha)}(x) = (-1)^n \frac{2^{-\alpha}}{\pi} \left(\frac{3}{2}\right)^{1/3} n_1^{-1/3} A_1(t) + O(n^{-1})$$

The asymptotic representation of $L_n^{(\alpha)}(x)$ in the second of the four regions, i.e. in the open interval $(0, 4n_1)$, may be derived by a different method (the *col* method); by this method it can be shown* that if

$$an_1 \leq x \leq bn_1$$

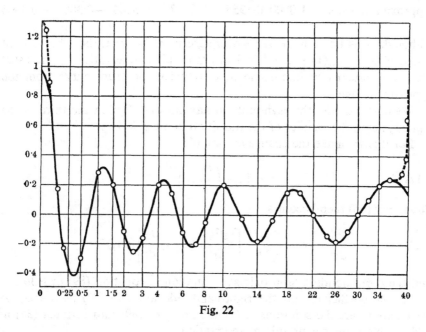

Fig. 22

where a and b are any two positive (fixed) numbers, then

(115) $$e^{-x/2} L_n^{(\alpha)}(x) = \frac{(-1)^n (2\cos\theta)^{-\alpha}}{\sqrt{(\pi n_1 \sin 2\theta)}} \left[\sin\Theta - \frac{A_1^{(\alpha)}(\theta) \cos\Theta}{n_1 \sin 2\theta} + O(n^{-2}) \right]$$

where $\quad \theta = \cos^{-1}\sqrt{\left(\frac{x}{4n_1}\right)}, \quad \Theta = n_1(2\theta - \sin 2\theta) + \frac{\pi}{4}$

$$A_1^{(\alpha)}(\theta) = \frac{1}{12}\left[\frac{5}{4\sin^2\theta} - (1 - 3\alpha^2)\sin^2\theta - 1\right]$$

This formula also gives excellent numerical results in general. In fact, the formula abbreviated to the single term in $\sin\Theta$ gives for $n = 10$, $\alpha = 0$,

---

* See the paper by F. TRICOMI cited on p. 186.

so good an approximation that the graph of the left-hand side is hardly distinguishable from that of the right-hand side. Figure 22 (which is actually drawn from a similar but *less exact* formula) shows both the exact and approximate values of $e^{-x/2} L_{10}(x)$ in the interval (0, 40), where for convenience the abscissæ are taken proportional to $\sqrt{x}$ instead of x. With the exception of the immediate vicinities of the extremes (at which the approximate curve is that shown by dots) the approximation is so close that only one graph has been drawn, that of the exact values, while some of the approximate values are indicated by circles.

39. The Legendre differential equation and polynomials

Another important example of problems of the type mentioned at the beginning of § 37 is *Legendre's equation*

(116) $$\frac{d}{dx}\left[(1-x^2)\frac{dy}{dx}\right]+\lambda y = 0$$

considered within the interval $(-1, 1)$, the extremes of which are singular points of the equation as the leading coefficient $p(x) = 1-x^2$ vanishes at these points. We look for *solutions which remain bounded as* $x \to +1$ *and* $x \to -1$.

As the coefficients of the *reduced* equation (i.e. with leading coefficient equal to 1) are analytical functions regular about the origin,* we try, as in § 37,

(117) $$y = c_0+c_1x+c_2x^2+\ldots$$

Thus

$$\frac{d}{dx}\left[(1-x^2)\frac{dy}{dx}\right] = 1\cdot 2c_2+2(3c_3-c_1)x+3(4c_4-2c_2)x^2+\ldots$$

and on substitution into (116) we obtain

$$\sum_{\nu=0}^{\infty}\{(\nu+1)[(\nu+2)c_{\nu+2}-\nu c_\nu]+\lambda c_\nu\}x^\nu = 0$$

---

* It will be established in the following chapter (§ 43) that *all* the integrals of a linear differential equation whose coefficients are analytic functions regular about a point x_0 (i.e. representable about x_0 as a power series in $x - x_0$) are analytic functions regular about $x = x_0$.

This immediately gives the recurrence relations between the coefficients

$$(118) \qquad c_{\nu+2} = \frac{\nu(\nu+1)-\lambda}{(\nu+1)(\nu+2)} c_\nu \qquad (\nu = 0, 1, 2, \ldots)$$

These formulæ suggest the separation in the power series (117) (which is permissible) of the terms of even and of odd degrees; we write (117) therefore in the form

$$(119) \qquad y = AF(x^2) + Bx\, G(x^2)$$

in which A and B are two arbitrary constants, and

$$(120) \qquad F(x^2) = \sum_{m=0}^{\infty} c_{2m} x^{2m}, \qquad G(x^2) = \sum_{m=0}^{\infty} c_{2m+1} x^{2m}$$

where the coefficients c_ν are determined by the preceding recurrence relations (118) and the *initial conditions* $c_0 = c_0^{(0)}$, $c_1 = c_1^{(0)}$.

It is at once clear that this process *does* in fact yield a solution, as the preceding series have non-zero radii of convergence—both series actually have *radius of convergence equal to* 1 as it follows immediately from (118) that

$$(121) \qquad \lim_{\nu \to \infty} \frac{c_\nu}{c_{\nu+2}} = 1$$

i.e. in both power series F and G the ratios of successive coefficients tend to 1 as $\nu \to \infty$.

We now discuss the behaviour of the functions F and G when x tends to either $+1$ or -1, i.e. when x^2 tends to 1. It will appear that exactly as was the case in § 37, if the power series considered do not terminate, so reducing to polynomials, the imposed conditions *cannot* be satisfied—in other words, we shall see that *excluding the case in which* λ *is of the form* $n(n+1)$ *where n denotes a non-negative integer, in which case either one or other of the two series* (120) *reduces to a polynomial* since at some point the numerator of the right-hand side of (118) becomes null, *both functions* $F(x^2)$ *and* $G(x^2)$ *tend to infinity as* $x^2 \to 1$.

For proof of this result, we observe in the first place that the relation (121) implies that from some point onwards the coefficients c_ν must have the same sign; we can suppose this sign to be positive in view of the arbitrariness of $c_0^{(0)}$ and $c_1^{(0)}$. We may therefore assume that for ν greater than a certain ν_0, $c_\nu > 0$.

§ 39] *The Legendre differential equation and polynomials*

In the second place, for $v > \lambda/2 - 1$,

$$\frac{v(v+1)-\lambda}{(v+1)(v+2)} > \frac{v-\lambda/2-1}{v-\lambda/2+1}$$

since on freeing of fractions, taking all terms to the left-hand side and simplifying, we obtain from this the inequality

$$2(v+1) + \tfrac{1}{2}\lambda^2 > 0$$

which is obviously satisfied. Further, since

$$\frac{v-\lambda/2-1}{v-\lambda/2+1} > \frac{v-r-1}{v-r+1}$$

provided that $v+1 > r > \lambda/2$, if we denote by $2r \geq v_0$ an even integer greater than $\lambda/2$ we find for $v > 2r$

$$\frac{c_{v+2}}{c_v} = \frac{v(v+1)-\lambda}{(v+1)(v+2)} > \frac{v-2r-1}{v-2r+1}$$

which implies

$$(v-2r+1)c_{v+2} > (v-2r-1)c_v$$

But this shows that the sequence of positive numbers

$$c_{2r+2}, \quad 3c_{2r+4}, \quad 5c_{2r+6}, \quad \ldots$$

is *increasing*, as is also the case for the sequence

$$2c_{2r+3}, \quad 4c_{2r+5}, \quad 6c_{2r+7}, \quad \ldots$$

Hence if ρ is the smaller of the two positive numbers c_{2r+2} and $2c_{2r+3}$,

(122) $$c_{2r+2s} \geq \frac{\rho}{2s-1}, \quad c_{2r+2s+1} \geq \frac{\rho}{2s} \quad (s = 1, 2, 3, \ldots)$$

Now suppose that the first of the two series has an infinite number of terms and write it as

$$F(x^2) = \sum_{m=0}^{r} c_{2m} x^{2m} + F^*(x^2), \quad \text{where} \quad F^*(x^2) = x^{2r} \sum_{s=1}^{\infty} c_{2r+2s} x^{2s}$$

The first part is simply a polynomial and in view of the first inequality in (122)

$$F^*(x^2) > x^{2r} \rho \sum_{s=1}^{\infty} \frac{x^{2s}}{2s-1}$$

But
$$\sum_{s=1}^{\infty} \frac{x^{2s}}{2s-1} = \frac{x}{2}\log\frac{1+x}{1-x}$$

whence
$$F^*(x^2) > \frac{\rho}{2}x^{2r+1}\log\frac{1+x}{1-x}$$

which shows that $F^*(x^2)$—and therefore also $F(x^2)$ which differs from F^* only by a polynomial—tends to *infinity* as $x \to \pm 1$.

The same argument may be repeated for $G(x^2)$; we have therefore shown that *the problem under consideration has a solution if and only if* λ *coincides with one of the* EIGENVALUES

(123) $$\lambda_n = n(n+1) \qquad (n = 0, 1, 2, \ldots)$$

and that *the corresponding* EIGENFUNCTION *(determined as far as a constant factor) is a polynomial of degree n to which either the function* $F(x^2)$ *or the function* $xG(x^2)$ *reduces, according as n is even or odd.*

To derive the eigenfunction explicitly we write $\lambda = n(n+1)$ in (118) and replace ν by $n-2h$, obtaining

$$\frac{c_{n-2h}}{c_{n-2h+2}} = -\frac{(n-2h+1)(n-2h+2)}{2h(2n-2h+1)}$$

Similarly
$$\frac{c_{n-2h+2}}{c_{n-2h+4}} = -\frac{(n-2h+3)(n-2h+4)}{2(h-1)(2n-2h+3)}$$

. . . .

$$\frac{c_{n-2}}{c_n} = -\frac{(n-1)n}{2\cdot 1(2n-1)}$$

and on multiplying together these h equalities we derive

$$c_{n-2h} = c_n(-1)^h \frac{n(n-1)\ldots(n-2h+1)}{2^h h!(2n-2h+1)(2n-2h+3)\ldots(2n-1)}$$

We now fix the values of the coefficients c_ν by choosing

$$c_n = \frac{1\cdot 3\cdot 5\ldots(2n-1)}{n!}$$

which implies
$$c_{n-2h} = (-1)^h \frac{n(n-1)\dots(n-2h+1) \, 1 \cdot 3 \cdot \dots \cdot (2n-2h-1)}{2^h h! \, n!}$$
$$= (-1)^h \frac{n(n-1)\dots(n-2h+1)}{2^h h! \, n!} \frac{(2n-2h)!}{2^{n-h}(n-h)!}$$
$$= \frac{(-1)^h}{2^n} \frac{n!}{h!(n-h)!} \frac{(2n-2h)!}{(n-2h)! \, n!} = \frac{(-1)^h}{2^n} \binom{n}{h} \binom{2n-2h}{n}$$

We therefore obtain the polynomial

(124) $$P_n(x) = \frac{1}{2^n} \sum_{h=0}^{[n/2]} (-1)^h \binom{n}{h} \binom{2n-2h}{n} x^{n-2h}$$

which is called the *Legendre polynomial of degree n*.

The first few of these polynomials which are of importance in many questions in both pure and applied mathematics are the following:

$$P_0(x) = 1, \quad P_1(x) = x, \quad P_2(x) = \tfrac{3}{2}x^2 - \tfrac{1}{2}$$
$$P_3(x) = \tfrac{5}{2}x^3 - \tfrac{3}{2}x, \dots$$

These polynomials contain only even or odd powers of x according as their suffix is even or odd.

40. An asymptotic expression for the Legendre polynomials

The Legendre polynomials possess many properties similar to those of the Laguerre polynomials and to those of eigenfunctions in general.

First, the *orthogonality relation*

(125) $$\int_{-1}^{+1} P_m(x) P_n(x) \, dx = 0 \qquad (m \neq n)$$

This follows, as we shall show, from (116). It should be noted that this relation is not immediately deducible from (61) of Chapter III on account of the singularities of the Legendre equation at the extremes of the fundamental interval.

For proof of (125) we observe that calculations exactly similar to those of § 26 yield the identity

$$[m(m+1) - n(n+1)] \int_{-1}^{+1} P_m(x) P_n(x) \, dx$$
$$= [(1-x^2)\{P_m(x) P_n'(x) - P_m'(x) P_n(x)\}]_{-1}^{+1}$$

As the function $p(x) = 1 - x^2$ vanishes for both $x = +1$ and for $x = -1$, while P_m, P_n, P_m' and P_n' remain finite at these two points, the relation (125) follows immediately.

It is a consequence of the orthogonality relation (125), by the general theorem proved in § 37, that *the zeros of $P_n(x)$ are all real, simple, and contained between -1 and $+1$.*

Other important properties of the Legendre polynomials are those expressed by the following formulæ similar to those stated for the Laguerre polynomials at the end of § 37:

(126) $$P_n(x) = \frac{1}{2^n n!} \frac{d^n}{dx^n}(x^2 - 1)^n$$

(127) $$(n+1)P_{n+1}(x) - (2n+1)x P_n(x) + n P_{n-1}(x) = 0$$

(128) $$(x^2 - 1)P_n'(x) = n[x P_n(x) - P_{n-1}(x)]$$

Also

(129) $$P_{2m+1}(0) = 0, \qquad P_{2m}(0) = (-1)^m \frac{1 \cdot 3 \cdot \ldots \cdot (2m-1)}{2 \cdot 4 \cdot \ldots \cdot (2m)}$$

$$P_n(1) = 1$$

For proof of these results we refer the reader to works on this particular subject.* Our interest here lies in the derivation of the important *asymptotic formula* (of Laplace) for the Legendre polynomials, viz.

(130) $$P_n(\cos\theta) = \sqrt{\left(\frac{2}{\pi n \sin\theta}\right)} \cos\left[(n+\tfrac{1}{2})\theta - \frac{\pi}{4}\right] + O(n^{-3/2})$$

valid in any interval contained in $(0, \pi)$, *the extremes excluded*. We use substantially the method of § 33 although the Legendre equation has singularities at the points $x = \pm 1$.

We begin by reducing the differential equation (116) to the form (68), by means of the transformations of page 165; here

$$p(x) = 1 - x^2, \qquad q(x) = 0, \qquad r(x) = 1$$

thus

$$\alpha(x) = (1-x^2)^{-1/4}, \qquad \xi = \int \frac{dx}{\sqrt{(1-x^2)}} = \sin^{-1} x$$

---

* See the works mentioned on p. 186. Books dealing with Legendre polynomials and related functions include J. Lense: *Kugelfunktionen* (Leipzig, Akad. Verlagsgesell., 1950) and E. W. Hobson: *Spherical and Ellipsoidal Harmonics* (Cambridge University Press, 1931). See also R. Lagrange (37).

An asymptotic expression for the Legendre polynomials

and therefore
$$q^* = p^{1/4}\frac{d}{dx}\left(p\frac{d\alpha}{dx}\right) = \frac{2-x^2}{4(1-x^2)} = \frac{1+\cos^2\xi}{4\cos^2\xi}$$

Hence, writing

(131) $\quad \lambda = \mu^2, \quad x = \sin\xi, \quad y = \frac{1}{\sqrt{(\cos\xi)}}z, \quad Q(\xi) = -\frac{1+\cos^2\xi}{4\cos^2\xi}$

the Legendre equation is transformed into the required form

(132) $\qquad \dfrac{d^2 z}{d\xi^2} + [\mu^2 - Q(\xi)]z = 0$

to be considered in the fundamental interval $(-\pi/2, +\pi/2)$.

It should be noticed that the coefficient $Q(\xi)$ has singularities at the extremes of the fundamental interval, and since

(133) $\quad \displaystyle\int Q(\xi)\,d\xi = -\tfrac{1}{4}\int\left(1+\frac{1}{\cos^2\xi}\right)d\xi = -\tfrac{1}{4}(\xi + \tan\xi)$

the existence of these singularities implies the *non-integrability of this co-efficient* in the fundamental interval, as $\tan\xi$ tends to infinity as $\xi \to \pm\pi/2$.

To avoid this difficulty it is convenient to cut off the extremes of the fundamental interval, i.e. to consider a variable ξ such that

(134) $\qquad -\dfrac{\pi}{2} + \varepsilon \leq \xi \leq \dfrac{\pi}{2} - \varepsilon$

where ε is a positive angle as small as we please (and certainly less than $\pi/2$).

We now employ the transformation of a differential equation into an integral equation as set out in § 29, to replace (132) by the integral equation

(135) $\quad z(\xi) = c_1 \cos\mu\xi + c_2 \sin\mu\xi + \dfrac{1}{\mu}\displaystyle\int_0^\xi Q(t)\sin\mu(\xi-t)\,z(t)\,dt$

where the two constants c_1 and c_2 are given by the formula

$$c_1 = z(0) = y(0), \quad \mu c_2 = z'(0) = \left\{\frac{d}{d\xi}[\sqrt{(\cos\xi)}\,y(\xi)]\right\}_{\xi=0} = y'(0)$$

Now identifying $y(\xi)$ with $P_n(\sin\xi)$ which implies

$$\mu = \sqrt{\lambda} = \sqrt{[n(n+1)]} = n\left(1+\frac{1}{n}\right)^{1/2} = n + \tfrac{1}{2} + O(n^{-1})$$

we deduce from (128) and the second relation (129)*

(if $n = 2m$)
$$\begin{cases} c_1 = P_{2m}(0) = (-1)^m \dfrac{1\cdot 3\cdots (2m-1)}{2\cdot 4\cdots (2m)} \\ \sqrt{[n(n+1)]}\, c_2 = P'_{2m}(0) = 2m P_{2m-1}(0) = 0 \end{cases}$$

(if $n = 2m+1$)
$$\begin{cases} c_1 = P_{2m+1}(0) = 0 \\ \sqrt{[n(n+1)]}\, c_2 = P'_{2m+1}(0) = (2m+1) P_{2m}(0) \\ \qquad = (-1)^m \dfrac{1\cdot 3\cdots (2m-1)}{2\cdot 4\cdots (2m)}(2m+1) \end{cases}$$

But

(136) $$\dfrac{1\cdot 3\cdots (2m-1)}{2\cdot 4\cdots (2m)} = \dfrac{1}{\sqrt{(\pi m)}} + O(m^{-3/2})$$

(as can be easily established**); consequently

---

* To avoid using the formulæ (128) and (129) which have not been established in this book we can calculate $P_n(0)$ and $P'_n(0)$ directly by putting $x = 0$ in the formulæ (124) and that obtained by differentiating (124) with respect to x.

** The easiest method of deriving (136) is to consider the integral

$$I_n = \int_0^{\pi/2} \cos^n x \, dx$$

which, as is at once established by integration by parts, satisfies the recurrence relation

$$I_n = \dfrac{n-1}{n} I_{n-2}, \qquad (n \geq 2)$$

Also $I_0 = \pi/2$ and $I_1 = 1$; whence

$$I_{2m} = \dfrac{1\cdot 3\cdots (2m-1)}{2\cdot 4\cdots (2m)} \dfrac{\pi}{2}, \qquad I_{2m+1} = \dfrac{2\cdot 4\cdots (2m)}{3\cdot 5\cdots (2m+1)}$$

or

$$I_{2m} = \dfrac{\pi}{2} R_m, \qquad I_{2m+1} = \dfrac{1}{(2m+1) R_m} = \dfrac{1}{(2m+2) R_{m+1}}$$

where R_m denotes $\dfrac{1\cdot 3\cdots (2m-1)}{2\cdot 4\cdots (2m)}$. Further

$$I_{2m-1} > I_{2m} > I_{2m+1};$$

therefore

$$\dfrac{1}{2m R_m} > \dfrac{\pi}{2} R_m > \dfrac{1}{(2m+1) R_m}$$

and this gives, on multiplying through by the positive quantity $2m R_m$,

$$1 > \pi m R_m^2 > 1 - \dfrac{1}{2m+1}$$

Consequently

$$\pi m R_m^2 = 1 + O(m^{-1})$$

from which follows (136), on dividing through by πm and taking the square root.

§40] An asymptotic expression for the Legendre polynomials

(137)
$$\begin{cases} (\text{if } n = 2m) & c_1 = \dfrac{(-1)^m}{\sqrt{(\pi m)}} + O(m^{-3/2}), \quad c_2 = 0 \\ (\text{if } n = 2m+1) & c_1 = 0, \quad c_2 = \dfrac{(-1)^m}{\sqrt{(\pi m)}} + O(m^{-3/2}) \end{cases}$$

as in the second case

$$c_2 = (-1)^m \frac{1 \cdot 3 \cdot \ldots \cdot (2m-1)}{2 \cdot 4 \cdot \ldots (2m)} \frac{2m+1}{2m + \tfrac{3}{2} + O(m^{-1})}$$

$$= (-1)^m \frac{1 \cdot 3 \cdot \ldots \cdot (2m-1)}{2 \cdot 4 \cdot \ldots (2m)} [1 + O(m^{-1})]$$

We now propose to show that *the function $z(\xi)$ is bounded in the interval* (134); in fact, we shall prove that if n is greater than a suitable n_0

(138) $$|z(\xi)| < 2$$

Within the interval defined in (134)

$$\left| \int_0^x |Q(\xi)| \, d\xi \right| = \tfrac{1}{4} |x + \tan x| \leq \tfrac{1}{4} \left[\frac{\pi}{2} - \varepsilon + \tan\left(\frac{\pi}{2} - \varepsilon\right) \right] < \frac{\pi}{8} + \tfrac{1}{4} \cot \varepsilon$$

Also, in view of (137), for μ sufficiently large (i.e. for n greater than a suitable n_0) either $|c_1| < 1$, $c_2 = 0$ or else $c_1 = 0$, $|c_2| < 1$, which implies in either case

$$|c_1 \cos \mu \xi + c_2 \sin \mu \xi| \leq \sqrt{(c_1^2 + c_2^2)} < 1$$

Hence if M is the greatest value of $|z(\xi)|$ in the interval (134) it follows from (135) that

$$M \leq 1 + \frac{1}{\mu} M \left| \int_0^\xi |Q(t)| \, dt \right| < 1 + M \frac{1}{\mu} \left(\frac{\pi}{8} + \tfrac{1}{4} \cot \varepsilon \right)$$

But by suitably increasing n_0 we may arrange that

$$\mu > 2 \left(\frac{\pi}{8} + \tfrac{1}{4} \cot \varepsilon \right)$$

so that
$$M < 1 + \tfrac{1}{2} M$$

which implies $M < 2$, i.e. (138) is established.*

---

* It can be shown by other methods that
$$|P_n(x)| \leq 1 \qquad (-1 \leq x \leq 1),$$
which implies $|z(\xi)| \leq 1$ within the interval $(-\pi/2, \pi/2)$, without any limitation on n.

This result implies that

$$\left|\int_0^\xi Q(t)\sin\mu(\xi-t)z(t)\,dt\right| < 2\left(\frac{\pi}{8}+\tfrac{1}{4}\cot\varepsilon\right)$$

and since μ is of the order of n,

$$z(\xi) = c_1\cos\mu\xi + c_2\sin\mu\xi + O(n^{-1})$$

i.e.

$$\sqrt{(\cos\xi)}\,P_n(\sin\xi) = c_1\cos\mu\xi + c_2\sin\mu\xi + O(n^{-1})$$

which, as we shall show, is essentially the Laplace formula.

Changing ξ into $\pi/2-\theta$ and remembering that $\mu = n+1/2+O(n^{-1})$, we obtain according as n is even ($n=2m$) or odd ($n=2m+1$)

$$\cos\mu\xi = \cos\left[(2m+\tfrac{1}{2})\frac{\pi}{2}-(2m+\tfrac{1}{2})\theta + O(n^{-1})\right] \quad (n\text{ even})$$

$$= \cos\left[m\pi+\frac{\pi}{4}-(2m+\tfrac{1}{2})\theta\right] + O(n^{-1})$$

$$= (-1)^m\cos\left[(2m+\tfrac{1}{2})\theta-\frac{\pi}{4}\right] + O(n^{-1})$$

$$\sin\mu\xi = \sin\left[(2m+\tfrac{1}{2})\frac{\pi}{2}+\frac{\pi}{2}-\left(2m+\tfrac{3}{2}\right)\theta + O(n^{-1})\right] \quad (n\text{ odd})$$

$$= \sin\left\{m\pi+\frac{\pi}{2}-\left[\left(2m+\tfrac{3}{2}\right)\theta-\frac{\pi}{4}\right]\right\} + O(n^{-1})$$

$$= (-1)^m\cos\left[\left(2m+\tfrac{3}{2}\right)\theta-\frac{\pi}{4}\right] + O(n^{-1})$$

Combining these results with (137) we obtain

$$\sqrt{(\sin\theta)}\,P_n(\cos\theta) = \begin{cases} \dfrac{1}{\sqrt{(\pi m)}}\cos\left[(2m+\tfrac{1}{2})\theta-\dfrac{\pi}{4}\right]+O(n^{-3/2}) & (n\text{ even}) \\ \dfrac{1}{\sqrt{(\pi m)}}\cos\left[\left(2m+\tfrac{3}{2}\right)\theta-\dfrac{\pi}{4}\right]+O(n^{-3/2}) & (n\text{ odd}) \end{cases}$$

and since, as far as the terms of order $O(n^{-3/2})$, we can in both cases identify $1/\sqrt{m}$ with $\sqrt{(2/n)}$ we have finally, for n even or odd,

$$\sqrt{(\sin\theta)}\,P_n(\cos\theta) = \sqrt{\left(\frac{2}{\pi n}\right)}\cos\left[(n+\tfrac{1}{2})\theta-\frac{\pi}{4}\right]+O(n^{-3/2})$$

which is exactly the Laplace formula (130).

§ 40] An asymptotic expression for the Legendre polynomials

In figure 23 are compared the graphs of $P_7(\cos \theta)$ (shown by the full line) and that of the right-hand side of the Laplace formula ($n = 7$) without residual term (shown by the dotted line). It is evident that despite the smallness of the value of n agreement between these curves is fairly close

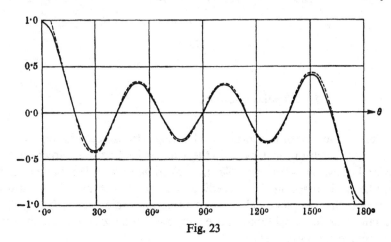

Fig. 23

except, as was to be expected, at the extremes*. This is further accentuated in the *diagram of errors* (figure 24) in which the difference between the value given by the Laplace formula and that of the exact value of $P_7(\cos \theta)$ is represented on a scale twenty-five times larger than that of figure 23.

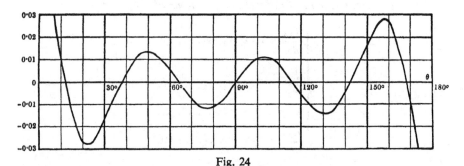

Fig. 24

This example, as also that of the Laguerre polynomials, justifies the earlier assertion that the asymptotic formulæ themselves generally furnish good approximations to the actual values of the functions.

---

* For an asymptotic representation of the Legendre polynomials (and, more generally, of the Jacobi polynomials) valid at the extremes of the fundamental interval ($-1, 1$), see F. TRICOMI: 'Expansions of the hypergeometric function in series of confluent ones and application to the Jacobi polynomials', *Comm. Math. Helvetici*, **25**, 196–204 (1951).

V.   Differential equations in the complex field

41. Majorizing functions

So far we have been concerned exclusively with differential equations *in the real field* which in most cases is the natural field to use. However there are several questions—for example, that of *asymptotic behaviour* which has already been discussed in the preceding chapter—which are most naturally developed in the complex field, or as is sometimes said the *analytic field*, as the functions under consideration are supposed to be *analytic*, i.e. they may be represented by power series.

As a first step in introducing the complex field we require to supplement the fundamental theorem of existence of Chapter I to establish that when the differential equations of a given system are analytic then the solutions are also analytic; we therefore prove the following theorem:

Given a normal system of differential equations of the form

(1)
$$\begin{cases} \dfrac{dy_1}{dx} = f_1(x, y_1, y_2, \ldots, y_n) \\ \dfrac{dy_2}{dx} = f_2(x, y_1, y_2, \ldots, y_n) \\ \ldots\ldots\ldots\ldots\ldots\ldots \\ \dfrac{dy_n}{dx} = f_n(x, y_1, y_2, \ldots, y_n) \end{cases}$$

where $f_1, f_2, \ldots, f_n$ are n analytic functions of the $n+1$ complex variables $x, y_1, y_2, \ldots, y_n$ regular within the vicinity of the point

(2) $x = x_0, \quad y_1 = y_1^{(0)}, \quad y_2 = y_2^{(0)}, \quad \ldots, \quad y_n = y_n^{(0)},$

i.e. such that each function considered as a function of x may be developed in a power series (of positive integral powers) of $x - x_0$, considered as a function of y_1 may be developed in a power series of $y_1 - y_1^{(0)}$, etc., where all the radii of convergence are non-zero, then there exists one and only one system of analytic functions of x regular within a certain vicinity (whose bounds can be explicitly stated), which system satisfies (1) identically and which assumes respectively the values $y_1^{(0)}$, $y_2^{(0)}$, ..., $y_n^{(0)}$ for $x = x_0$.

This theorem may be proved by the method of successive approximations as used in Chapter I, with few modifications. It is however of real value to use a different method—a method which may also be used for systems of equations in the real field although it belongs most naturally to the complex field—originally used by Cauchy and developed earlier than the method of successive approximations; it was in fact the first method to produce rigorous results in this branch of analysis.

The method of Cauchy hinges upon the idea of *majorizing functions*; in order to understand this notion fully we first consider an analytic function $f(x)$ of one complex variable x regular in the neighbourhood of some point x_0 in the complex plane. We say that another analytic function $\phi(x)$ majorizes $f(x)$ (in the neighbourhood of the point x_0) and we write

$$\phi(x) \gg f(x), \quad \text{or} \quad f(x) \ll \phi(x)$$

if for all integral values of $n \geq 0$

(3) $$\phi^{(n)}(x_0) \geq |f^{(n)}(x_0)|$$

(This definition implies that all the derivatives of $\phi(x)$ at the point x_0 are real non-negative numbers.)

A majorizing function for a given function $f(x)$ is easily obtained*; for if M denotes the maximum of $|f(x)|$ in a circle of radius r with centre at x_0 within which the function is regular, then**

(4) $$|f^{(n)}(x_0)| \leq M \frac{n!}{r^n} \qquad (n = 0, 1, 2, \ldots)$$

Now the analytic function

(5) $$\phi(x) = \frac{M}{1 - \frac{x - x_0}{r}}$$

which within the circle $|x - x_0| < r$ is identical with the sum of the geometric series

$$M \sum_{n=0}^{\infty} \left(\frac{x - x_0}{r} \right)^n = \sum_{n=0}^{\infty} M \frac{n!}{r^n} \cdot \frac{(x - x_0)^n}{n!}$$

---

* It is clear that there is an infinity of majorizing functions of a given function $f(x)$; for if $\phi_0(x)$ is any majorizing function of $f(x)$ then so also is $\phi_0(x) + \psi(x)$ where $\psi(x)$ is any analytic function whose derivatives are all real and *non-negative* at the point $x = x_0$.
** See, for example, F. TRICOMI (80).

has as its nth derivative at the point $x = x_0$ the term on the right-hand side of (4); whence

$$|f^{(n)}(x_0)| \leq \phi^{(n)}(x_0)$$

i.e.

$$f(x) \ll \phi(x)$$

If, at the point x_0, $f(x) \ll \phi(x)$ and if in a neighbourhood of this point the function $\phi(x)$ may be expanded as a power series in $x - x_0$, so also may the function $f(x)$ and consequently

$$|f(x)| \leq |f(x_0)| + |f'(x_0)|\frac{|x-x_0|}{1!} + |f''(x_0)|\frac{|x-x_0|^2}{2!} + \cdots$$

$$\leq \phi(x_0) + \phi'(x_0)\frac{|x-x_0|}{1!} + \phi''(x_0)\frac{|x-x_0|^2}{2!} + \cdots = \phi(x_0 + |x-x_0|)$$

Hence, in particular for $x_0 = 0$,

$$|f(x)| \leq \phi(|x|)$$

The idea of majorizing functions may be extended in an obvious way to cover functions of more than one independent variable; for example, for the function $f(x, y)$ of the two independent variables x and y, analytic and regular within the domain $|x - x_0| < r$ (say, the circle c in the x-plane), $|y - y_0| < r'$ (say, the circle c' in the y-plane) and satisfying $|f(x, y)| \leq M$, if we put

(6)
$$\phi(x, y) = \frac{M}{\left(1 - \dfrac{x-x_0}{r}\right)\left(1 - \dfrac{y-y_0}{r'}\right)}$$

then

$$f(x, y) \ll \phi(x, y)$$

in the sense that for any non-negative integers m and n

(7)
$$\left|\frac{\partial^{m+n}f}{\partial x^m \partial y^n}\right| \leq \frac{\partial^{m+n}\phi}{\partial x^m \partial y^n} \quad \text{at the point } (x_0, y_0)$$

The Cauchy formula for the mth derivative* of $\partial^n f/\partial y^n$ at x_0 within the circle c in the x-plane yields

$$\left(\frac{\partial^{m+n}f}{\partial x^m \partial y^n}\right)_{\substack{x=x_0\\y=y_0}} = \frac{m!}{2\pi i}\int_c \frac{1}{(x-x_0)^{m+1}}\left(\frac{\partial^n f}{\partial y^n}\right)_{y=y_0} dx$$

---

* See, for example, F. TRICOMI (80), p. 41.

But the inequality (4) applied to the nth derivative of f with respect to y yields

$$\left|\left(\frac{\partial^n f}{\partial y^n}\right)_{y=y_0}\right| \leq M \frac{n!}{r'^n}$$

and it therefore follows from the last equation, by taking moduli on both sides, that

$$\left|\left(\frac{\partial^{m+n} f}{\partial x^m \partial y^n}\right)_{\substack{x=x_0 \\ y=y_0}}\right| \leq \frac{m!}{2\pi} \int_c \frac{1}{r^{m+1}} M \frac{n!}{r'^n} |dx| = \frac{M m! n!}{r^m r'^n}$$

On the other hand, within the domain $c+c'$,

$$\phi(x, y) = M \sum_{m=0}^{\infty} \left(\frac{x-x_0}{r}\right)^m \sum_{n=0}^{\infty} \left(\frac{y-y_0}{r'}\right)^n = \sum_{m,n=0}^{\infty} \frac{M m! n! (x-x_0)^m (y-y_0)^n}{r^m r'^n \, m! \, n!}$$

whence

$$\left(\frac{\partial^{m+n} \phi}{\partial x^m \partial y^n}\right)_{\substack{x=x_0 \\ y=y_0}} = \frac{M m! n!}{r^m r'^n}$$

Therefore for the function $\phi(x, y)$ defined by (6) the inequality (7) is valid, i.e. $\phi(x, y) \geqslant f(x, y)$.

The further extension to the case of three or more variables is done in an exactly similar way.

It should be noted that a majorizing function is not necessarily a real function, but must necessarily assume real non-negative values, as must also its derivatives, at the particular point at which it majorizes the given function.

42. Proof of the fundamental theorem by Cauchy's method

We now present a proof by Cauchy's method of the fundamental theorem stated in the last section—a method which he called *the method of limits* but which today is more appropriately called *the method of majorizing functions*.

We write the equations (1) in the more concise form

(8) $$\frac{dy_h}{dx} = f_h(x, y_1, y_2, \ldots, y_n) \qquad (h = 1, 2, \ldots, n)$$

and the initial conditions (2) in the form

(9) $$y_h(0) = 0 \qquad (h = 1, 2, \ldots, n)$$

where we have supposed the point $(x_0, y_1^{(0)}, \ldots, y_n^{(0)})$ to be the origin. As the functions f_h may be differentiated as often as we please we can write

(10)
$$\begin{cases} y_h'(0) = (f_h)_0, \quad y_h''(0) = \left(\frac{\partial f_h}{\partial x}\right)_0 + \sum_{k=1}^n \left(\frac{\partial f_h}{\partial y_k}\right)_0 (f_k)_0 \\ y_h'''(0) = \left(\frac{\partial^2 f_h}{\partial x^2}\right)_0 + \sum_{k=1}^n \left[2\left(\frac{\partial^2 f_h}{\partial x \partial y_k}\right)_0 (f_k)_0 + \left(\frac{\partial f_h}{\partial y_k}\right)_0 \left(\frac{\partial f_k}{\partial x}\right)_0 \right] \\ \qquad + \sum_{k,l=1}^n \left[\left(\frac{\partial^2 f_h}{\partial y_k \partial y_l}\right)_0 (f_k)_0 (f_l)_0 + \left(\frac{\partial f_h}{\partial y_k}\right)_0 \left(\frac{\partial f_k}{\partial y_l}\right)_0 (f_l)_0 \right] \\ \qquad \cdots \cdots \cdots \cdots \cdots \cdots \cdots \cdots \cdots \cdots \cdots \end{cases}$$

where the suffix 0 denotes that the function or its derivatives, as may be, is to be evaluated at the origin, i.e. where $x = y_1 = y_2 = \ldots = y_n = 0$. It is important to note that $y_h^{(n)}$ is always a polynomial (of degree n) in the functions f_k and their derivatives of orders up to $n-1$, whose coefficients are all positive numbers.

If therefore the system (8) with the initial conditions (9) possesses an analytic solution regular in the vicinity of $x = 0$ it must be given by

(11) $$y_h(x) = y_h'(0)\frac{x}{1!} + y_h''(0)\frac{x^2}{2!} + y_h'''(0)\frac{x^3}{3!} + \ldots \qquad (h = 1, 2, \ldots, n)$$

where $y_h'(0), y_h''(0), \ldots$ assume the values given by (10). We now consider whether this *can* be a solution or not, i.e. whether these series (11) have non-zero radii of convergence.

To investigate this question we use *the method of majorizing functions*, the basic idea being *to consider in place of the system (8)–(9) the other system*

(12) $$\frac{dY_h}{dx} = F_h(x, Y_1, Y_2, \ldots, Y_n), \qquad Y_h(0) = 0 \qquad (h = 1, 2, \ldots, n)$$

where F_h denotes a function which majorizes f_h; we shall prove that *a solution of this system provides a n-ple of functions which majorize the functions y_h whose derivatives are given by* (10). It will immediately follow from this that if the solution of (12) under consideration may be represented by a power series in x convergent within a certain circle c in the x-plane with centre the point $x = 0$, then the series (11) must converge within the same circle, for the series of moduli of the terms in (11) are majorized* by the similar series of moduli of the terms in (12).

To prove the theorem stated above we need only note that the values of $Y_h'(0)$, $Y_h''(0)$, $Y_h'''(0)$, ... are calculated by formulæ exactly similar to (10) but in which the f's are replaced by the corresponding F's; since $F_h \gg f_h$ we derive successively

$$|y_h'(0)| = |(f_h)_0| \leq F_h(0) = Y_h'(0)$$

$$|y_h''(0)| \leq \left|\left(\frac{\partial f_h}{\partial x}\right)_0\right| + \sum_{k=1}^{n}\left|\left(\frac{\partial f_h}{\partial y_k}\right)_0\right||(f_k)_0| \leq \left(\frac{\partial F_h}{\partial x}\right) + \sum_{k=1}^{n}\left(\frac{\partial F_h}{\partial y_k}\right)(F_k)_0 = Y_h''(0)$$

. . . .

whence follow the inequalities

$$|y_h(0)| = Y_h(0), \quad |y_h'(0)| \leq Y_h'(0), \quad |y_h''(0)| \leq Y_h''(0), \quad \ldots \quad (h = 1, 2, \ldots, n)$$

Therefore $y_h(x) \ll Y_h(x)$ at the origin.

Now suppose that within the domain $|x| < a$, $|y_h| < b$ ($h = 1, 2, 3, \ldots, n$) the n analytic functions $f_h(x, y_1, y_2, \ldots, y_n)$ are regular; there therefore exists a positive constant M such that

$$|f_h(x, y_1, y_2, \ldots, y_n)| < M \qquad (h = 1, 2, \ldots, n)$$

Consequently, as shown in the preceding section, the function

$$F(x, y_1, y_2, \ldots, y_n) = \frac{M}{\left(1 - \frac{x}{a}\right)\left(1 - \frac{y_1}{b}\right)\left(1 - \frac{y_2}{b}\right) \cdots \left(1 - \frac{y_n}{b}\right)}$$

majorizes each of the n functions f_h, and the system (12) may be taken as

$$\frac{dY_h}{dx} = \frac{M}{\left(1 - \frac{x}{a}\right)\left(1 - \frac{Y_1}{b}\right)\left(1 - \frac{Y_2}{b}\right) \cdots \left(1 - \frac{Y_n}{b}\right)}, \quad Y_h(0) = 0 \quad (h = 1, 2, \ldots, n)$$

This implies that

$$Y_1(x) \equiv Y_2(x) \equiv Y_3(x) \equiv \ldots \equiv Y_n(x)$$

---

* The series of positive terms Σa_n is said to *majorize* the series Σb_n, also of positive terms, if for any n we have $a_n > b_n$. Therefore, if the first series is convergent, so also is the second.

Hence if $Y(x)$ is the common value of the n functions $Y_h(x)$, $Y(x)$ is determined from the equation

$$\frac{dY}{dx} = \frac{M}{\left(1-\frac{x}{a}\right)\left(1-\frac{Y}{b}\right)^n}$$

i.e.

$$\left(1-\frac{Y}{b}\right)^n dY = M \frac{dx}{\left(1-\frac{x}{a}\right)}$$

which gives, on integrating,

$$-\frac{b}{(n+1)}\left(1-\frac{Y}{b}\right)^{n+1} = -aM \log\left(1-\frac{x}{a}\right) + \text{constant}$$

and therefore

(13) $$Y(x) = b\left\{1 - \left[1 + \frac{(n+1)aM}{b}\log\left(1-\frac{x}{a}\right)\right]^{\frac{1}{n+1}}\right\}$$

where the logarithm takes *its principal value** and the constant of integration has been chosen so that $Y = 0$ for $x = 0$.

It therefore follows that the series (11) is certainly convergent whenever the series for $Y(x)$ given by (13) converges. But this function has singular points only at $x = a$ and at the point x_0 for which

$$\frac{(n+1)aM}{b}\log\left(1-\frac{x_0}{a}\right) = -1$$

i.e. the point

$$x_0 = a(1 - e^{-\frac{b}{(n+1)aM}})$$

Of these two singular points the one nearer to the origin is the point x_0; the series (11) is therefore convergent at least within the circle with centre the origin and radius

(14) $$r = a(1 - e^{-\frac{b}{(n+1)aM}})$$

In other words, *the radius of convergence of the series* (11) *is not less than the positive number r given by the formula* (14).

---

* That value which reduces to a real number when the argument is real and positive.

This completes the proof of the fundamental theorem stated in the last section, and in fact we have even specified a minimum radius of convergence about the origin. Within the circle of this radius about the origin the solution of (8) whose existence is established by the fundamental theorem is analytic.

It should be noted that the uniqueness of the solution is established *in the field of analytic functions regular about the origin*. It is still doubtful whether there can also exist other solutions which are not regular about the origin, for example solutions of the type $x^{\alpha}f(x)$ where $f(x)$ is an analytic function regular about the origin and α a positive *non-integral* number. To settle this question further considerations are necessary; these may be based either on the method of majorizing functions* or, as is simpler, on the method of successive approximations of Chapter I now used for the complex field. By this method the existence and uniqueness of the solution can be established without the necessity of postulating analyticity.

In view of the importance of the result it is worth while to set out the fundamental theorem for the case of *the single equation of second order*

(15) $$y'' = F(x, y, y')$$

together with the initial conditions

(16) $$y(x_0) = y_0, \qquad y'(x_0) = y'_0$$

The corresponding normal system takes the form

$$\begin{cases} \dfrac{dy_1}{dx} = y_2 \\ \dfrac{dy_2}{dx} = F(x, y_1, y_2) \end{cases} \qquad \begin{cases} y_1(x_0) = y_0 \\ y_2(x_0) = y'_0 \end{cases}$$

The fundamental theorem is as follows:

If the function $F(x, y, y')$ is an analytic function of its three arguments, regular within the domain

(17) $$|x - x_0| \leq a, \qquad |y - y_0| \leq b, \qquad |y' - y'_0| \leq b$$

and if in this domain

$$|F(x, y, y')| \leq N$$

equation (15) *has one and only one solution satisfying the initial conditions* (16),

---

* See, for example, E. PICARD (74), Vol. II, Chapter XI, § 4, or, preferably, L. BIEBERBACH, (4), p. 10.

this solution being an analytic function of x regular at least within the circle c with centre x_0 and radius

(18) $$r = a(1 - e^{-b/3aM})$$

where M is the greater of the two numbers N and $|y'_0| + b$.*

The function $Y(x)$ given by (13) (with $n = 2$) majorizes both $y \equiv y_1$ and $y' \equiv y_2$ at the origin; hence, within the circle c,

(19) $$|y - y_0| \leq Y(|x - x_0|), \qquad |y' - y'_0| \leq Y(|x - x_0|)$$

where

(20) $$Y(x) = b \left\{ 1 - \left[1 + \frac{3aM}{b} \log\left(1 - \frac{x}{a}\right) \right]^{\frac{1}{3}} \right\}$$

As it is obvious that within c the quantity $Y(|x - x_0|) < b$, we deduce from (20) that

$$|y - y_0| < b, \qquad |y' - y'_0| < b$$

thus eliminating any possibility that the solution may pass out of the domain (17).

43. General remarks on singular points of solutions of differential equations. The case of linear equations

The results so far obtained ensure that given a differential equation of the form

$$y^{(n)} = F(x, y, y', \ldots, y^{(n-1)})$$

in which F is an analytic function regular at the point

$$x = x_0, \qquad y = y_0, \qquad y' = y'_0, \qquad \ldots, \qquad y^{(n-1)} = y_0^{(n-1)}$$

then its solution which satisfies the initial conditions

$$y(x_0) = y_0, \qquad y'(x_0) = y'_0, \qquad \ldots, \qquad y^{(n-1)}(x_0) = y_0^{(n-1)}$$

is an analytic function *regular in a suitable neighbourhood of the point* x_0. We ask how this solution behaves away from the point x_0.

---

* M is such that
$$|y_2| \leq M, \quad \text{and} \quad |F(x, y_1, y_2)| \leq M,$$
since
$$|y_2| = |y'| = |y'_0 + (y' - y'_0)| \leq |y'_0| + |y' - y'_0| \leq |y'_0| + b$$

§ 43] Singular points of solutions of differential equations

Some simple examples suffice to show that away from the point x_0 singularities may occur; these may be *movable singularities* (whose position depends on the initial conditions) of the integrals of the equation although not singularities of the equation itself. For example, the simple equation

(21) $$y' + y^2 = 0$$

has general integral (see page 13)

(22) $$y = \frac{1}{x-c}$$

which therefore possesses a *pole* (of first order) at the point $x = c$, a variable point determined by the arbitrary constant c.

However this complication does not arise for some important classes of differential equations, those with *fixed singularities*; included among these are linear differential equations. For this last class of equations we shall establish the following important theorem:

The only possible finite singular points of the integrals of a linear differential equation whose leading coefficient* equals one and whose other coefficients are analytic functions of the independent variable are the singular points of the coefficients themselves.

For example, for the linear equation of *second* order

(23) $$y'' + p_1(x)y' + p_2(x)y = 0$$

in which $p_1(x)$ and $p_2(x)$ are analytic functions of x, the singular points of $p_1(x)$ and $p_2(x)$ are the *only possible singular points of any integral of the equation*; i.e. if $x = x_0$ is an ordinary point of both $p_1(x)$ and $p_2(x)$ we may construct a circle (of non-zero radius) with centre at x_0 within which *all* the integrals of (23) are regular.

Let y_0 and y_0' denote the values assumed at $x = x_0$ by some integral y of (23) and its first derivative, and a a positive number such that both $p_1(x)$ and $p_2(x)$ are regular for $|x - x_0| \leq a$; since the function

$$F(x, y, y') = -p_1(x)y' - p_2(x)y$$

is regular within the bounds for y and y', it follows from the remarks at the end of the previous section that given b, any positive number, and N the maximum modulus of $F(x, y, y')$ in the domain

(24) $$|x - x_0| \leq a, \quad |y - y_0| \leq b, \quad |y' - y_0'| \leq b$$

---
* The coefficient of the derivative of highest order.

the function $y(x)$ is regular within the circle with centre at x_0 and radius

(25) $$r = a(1 - e^{-b/3aM})$$

M being the greater of the two numbers N and $|y_0'| + b$ (or a number greater than either of these).

This however is not sufficient to establish the theorem since, although b may possibly be chosen as we please, it cannot be known *a priori* whether the lower bound of values of r (as y_0 and y_0' vary) is non-zero or zero,* and in the latter case we cannot assert that *all* the integrals of the equation are regular about the point x_0.

We deal with this difficulty as follows; let P_1 and P_2 denote the maximum moduli of $p_1(x)$ and $p_2(x)$ respectively within the circle $|x - x_0| \leq a$; then within the domain (24)

$$|-p_1(x)y' - p_2(x)y| \leq P_1(|y_0'| + b) + P_2(|y_0| + b)$$

Hence on writing

$$\max(1, P_1 + P_2) = P, \quad \max(|y_0'|, P_1|y_0'| + P_2|y_0|) = Y$$

we may assume**

$$M = Pb + Y$$

which, by (25), implies

$$r = a\left(1 - e^{-\frac{1}{3a(P + Y/b)}}\right)$$

---

* It is easily seen that this in fact happens for equation (21), in which $n = 1$, $M = N = (|y_0| + b)^2$, and therefore
$$r = a[1 - e^{-b/2a(|y_0| + b)^2}]$$
where a and b are any two positive numbers. Writing for convenience $\xi = b/2a(|y_0| + b)^2$ we derive
$$r = a(1 - e^{-\xi}) < \lim_{a \to +\infty} r = \frac{b}{2(|y_0| + b)^2}$$
as
$$\frac{\partial r}{\partial a} = 1 - (1 + \xi)e^{-\xi} > 0$$
This yields
$$r < \frac{1}{8|y_0|}$$
since
$$\frac{\partial}{\partial b} \frac{b}{2(|y_0| + b)^2} = \frac{|y_0| - b}{2(|y_0| + b)^3}$$
and the function $b/2(|y_0| + b)^2$ therefore has a *maximum* value (equal to $1/8|y_0|$) for $b = |y_0|$. But the ratio $1/8|y_0|$ can be made as small as we please by choosing $|y_0|$ sufficiently large; hence its lower bound, and therefore that of the values of r, is zero.

** For M is the greater of the two numbers $P_1(|y_0'| + b) + P_2(|y_0| + b)$ and $|y_0'| + b$; consequently we cannot assert that M is greater than the sum of the maximum of the two quantities $\{P_1|y_0'| + P_2|y_0|\}$ and $|y_0'|$ plus b times the maximum of the two numbers $P_1 + P_2$ and 1.

The right-hand side above is obviously a *decreasing* function of Y/b; hence, since in view of the arbitrariness of b we may arrange that $Y/b \leq P$, i.e. $b \geq Y/P$, the value of the right-hand side is not less than that derived by putting $Y/b = P$. Therefore, by suitably choosing the number b, we can ensure that $r \geq \rho$ where

$$\rho = a(1 - e^{-1/6aP}) \tag{26}$$

It follows that the lower bound of values of r is, at worst, equal to the positive number ρ defined above, and the theorem is established.

We repeat that the singular points of the coefficients *may* be (not *must* be) singular points of the integrals of the equation. For example, in the case of the equation

$$y'' + \frac{h}{x} y' + \frac{k}{x^2} y = 0 \tag{27}$$

where h and k are constants (and therefore in which the coefficients are singular at the origin) the general integral is given by

$$y = c_1 x^{r_1} + c_2 x^{r_2} \tag{28}$$

where r_1 and r_2 denote the roots of the quadratic equation

$$r(r-1) + hr + k = 0 \tag{29}$$

as can be seen on substituting $y = x^r$ into the differential equation; the integrals will be singular or non-singular at the origin according to the values of h and k. Thus, for example, if $h = -2$, $k = 2$, in which case $r_1 = 1$, $r_2 = 2$, *no* integral of the equation is singular at the origin; while if instead $h = 4$, $k = 2$ in which case $r_1 = -1$, $r_2 = -2$, *all* the integrals are singular at the point $x = 0$. Finally, if $h = 1$, $k = -1$, giving $r_1 = 1$, $r_2 = -1$, the ∞^1 integrals of the type $c_1 x^{r_1}$ are regular (in fact, are zero) at the origin, while all other integrals are singular there.

The last case illustrates the fact that even for linear equations the singularities may depend in some way on the initial conditions, but only in the sense that the singular points of the coefficients may be singular points of some integrals of the equation and not of others.

We note in conclusion that the singular points of the coefficients— which we shall always suppose to be *isolated* points—may be either *poles* or *essential singularities* or else *branch points** of the integrals of the equations, as, for example, occurs for the equation (27) when the roots r_1 and r_2 of (29) are not both integral.

---

* The branch points of a multivalent function are the points about which the various values of the function permute. See, for example, F. TRICOMI (80), p. 91.

44. Investigation of the many-valuedness of integrals of a linear equation

The fact which we have already stressed that the singular points of the coefficients of a linear differential equation may be branch points of some of its integrals, i.e. that these integrals may be *non-uniform* analytic functions, although the coefficients in the equation are uniform, deserves particular attention as in analysis the representation of non-uniform functions always presents serious complications and may sometimes lead to ambiguity and error. Fortunately however, the possible many-valuedness of these integrals is fairly easily dealt with as it arises from factors or terms that can be explicitly determined while the functions which in general cannot be explicitly found remain uniform.

For simplicity we again consider a linear equation of the second order of the form (23) and try to determine the behaviour of two linearly independent integrals $y_1(x)$ and $y_2(x)$ of this equation in the vicinity of the point x_0 in the complex x-plane, where x_0 is a singular point for at least one of the two coefficients of the equation. We consider a variable point x initially at the point $x = x^*$ within a neighbourhood σ of the point x_0 all of whose points (with the exception of x_0) are ordinary points* of the coefficients in the equation, and then tracing out a closed curve Γ (for example, a circle with centre x_0) which lies within σ and which encircles exactly once (in the positive direction with respect to the area enclosed by it) the singular point x_0. As x varies along the curve Γ the two power series in $x - x^*$ which originally represented (within some suitable circle with centre x^*) the two linearly independent integrals $y_1(x)$ and $y_2(x)$ are said to be *analytically continued* along the curve Γ; these two power series when analytically continued along the entire closed curve Γ will give rise to two new power series in $x - x^*$ which we denote by $Y_1(x)$ and $Y_2(x)$.

When the two integrals $y_1(x)$ and $y_2(x)$ are known to be *uniform* analytic functions within the domain σ then evidently $Y_1(x) \equiv y_1(x)$, $Y_2(x) \equiv y_2(x)$; if however it is not known *a priori* that this is the case, we must in general expect that y_1 and y_2 are altered. All that we can assert is that *there exist four constants a, b, c, d such that*

$$(30) \qquad \begin{vmatrix} a & b \\ c & d \end{vmatrix} \neq 0$$

---

* That such a neighbourhood exists follows from the earlier hypothesis that the singular points are *isolated*.

§ 44] *Investigation of the many-valuedness of integrals of a linear equation* 215

where

(31) $\quad Y_1(x) \equiv ay_1(x) + by_2(x), \quad Y_2(x) \equiv cy_1(x) + dy_2(x)$

It is clear that Y_1 and Y_2 are integrals of (23)—and therefore expressible as suitable linear combinations of the two linearly independent integrals y_1 and y_2—as at all stages of the analytical continuation along the curve Γ equation (23) remains identically satisfied. Also, the condition (30) must be satisfied as otherwise the integrals $Y_1(x)$ and $Y_2(x)$ would not be linearly independent which would imply that $y_1(x)$ and $y_2(x)$ would not be linearly independent—for a relation $\alpha Y_1(x) + \beta Y_2(x) = 0$ on analytical continuation in the reverse direction along the closed curve Γ gives rise to a similar relation between y_1 and y_2.*

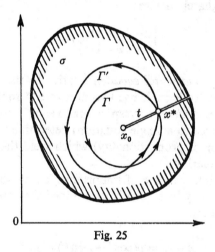

Fig. 25

We now consider the behaviour of another integral y_0 of the equation, by expressing it in terms of the fundamental system formed by y_1 and y_2. We write

$$y_0 = k_1 y_1 + k_2 y_2$$

where k_1 and k_2 are constants; let Y_0 be the integral obtained by analytically continuing the integral y_0 along the closed curve Γ; then

(32) $\quad Y_0 = k_1 Y_1 + k_2 Y_2 = (k_1 a + k_2 c) y_1 + (k_1 b + k_2 d) y_2$

---

* The constants a, b, c, d depend in general on the fundamental system y_1 and y_2 but not on the particular cycle Γ considered; for if, in addition to the cycle Γ we consider the second cycle Γ' similar to Γ (cf. figure 25), the linear substitution corresponding to (31) relative to Γ' must be (31) itself, for on traversing first the cycle Γ in the positive sense and then the cycle Γ' in the negative sense we must finish at the point x^* with the same y_1 and y_2 as we began with, since the integrals considered are uniform in the domain σ with a suitable *crosscut*, for example a crosscut taken along the line t as indicated in figure 25.

From this it follows immediately that the necessary and sufficient condition that a certain integral y should be *invariant* with respect to the closed curve Γ, i.e. that it should alter only by a factor of proportionality λ on circuit around Γ, is that the equations

$$k_1 a + k_2 c = \lambda k_1, \qquad k_1 b + k_2 d = \lambda k_2$$

are simultaneously satisfied, i.e. that k_1 and k_2 satisfy the two linear homogeneous equations

(33)
$$\begin{cases} (a-\lambda)k_1 + ck_2 = 0 \\ bk_1 + (d-\lambda)k_2 = 0 \end{cases}$$

From this we immediately derive

(34)
$$\begin{vmatrix} a-\lambda & c \\ b & d-\lambda \end{vmatrix} = 0$$

which is called the *fundamental equation* relative to the singular point* x_0. This equation, a quadratic in λ, has two non-zero** roots λ_1 and λ_2 which meantime we suppose *distinct*; corresponding to each of these roots equations (33) (which now reduce to a single equation) determine the values of k_1 and k_2 (as far as a factor of proportionality) which yield the invariant integrals of the differential equation.

These invariant integrals are of importance, as we shall see, since apart from a factor which may be explicitly determined they are *uniform* functions of x about the point x_0.

Putting

$$x - x_0 = \rho(\cos\theta + i\sin\theta) = \rho e^{i\theta}$$

we have

$$(x-x_0)^r = \rho^r e^{ri\theta}$$

Hence the function $(x-x_0)^r$ (which is non-uniform if r is not integral) becomes multiplied by $e^{2\pi i r}$ on describing a circuit around x_0, i.e. is multiplied by λ if r is chosen so that

$$e^{2\pi i r} = \lambda$$

i.e.

(35)
$$r = \frac{1}{2\pi i}\log\lambda$$

---

* It can be easily shown that while a, b, c, d depend on the particular fundamental system y_1, y_2 considered, the roots of equation (34) are independent of the fundamental system used. They depend only on the coefficients in the differential equation and on x_0.

** $\lambda_1 \lambda_2 = ad - bc \neq 0$, by (30).

It follows that if

(35') $$r_1 = \frac{1}{2\pi i}\log\lambda_1, \qquad r_2 = \frac{1}{2\pi i}\log\lambda_2$$

and if $\bar{y}_1$ and $\bar{y}_2$ be two given invariant integrals of (23) with multipliers λ_1 and λ_2, the two functions

$$\phi_1(x) = \frac{\bar{y}_1(x)}{(x-x_0)^{r_1}}, \qquad \phi_2(x) = \frac{\bar{y}_2(x)}{(x-x_0)^{r_2}}$$

are *unaltered* on analytical continuation around x_0 and therefore are analytic functions *uniform* within the neighbourhood σ about this point.

It may therefore be concluded that *when the fundamental equation* (34) *relative to the singular point* x_0 *has two distinct roots* λ_1 *and* λ_2 *the differential equation* (23) *possesses two linearly independent integrals which may be represented in the neighbourhood of the singular point* x_0 *by*

(36) $$y_1(x) = (x-x_0)^{r_1}\phi_1(x), \qquad y_2(x) = (x-x_0)^{r_2}\phi_2(x)$$

where the exponents r_1 and r_2 are given by (35') and $\phi_1(x)$ and $\phi_2(x)$ are analytic functions uniform about the point x_0.

It is clear therefore that the possible many-valuedness* of the two integrals depends on the explicit factors $(x-x_0)^{r_1}$ and $(x-x_0)^{r_2}$ alone.

We now consider the case in which the fundamental equation (34) has a *double* root $\lambda_1 = \lambda_2$.

In this case there is a single invariant integral $\bar{y}_1$ (with multiplier λ_1) with which we associate a second integral y_2 linearly independent of $\bar{y}_1$; the substitution (31) here assumes the form

$$Y_1 = \lambda_1 y_1, \qquad Y_2 = cy_1 + dy_2$$

or

(37) $$Y_1 = \lambda_1 y_1, \qquad Y_2 = cy_1 + \lambda_1 y_2$$

since $d = \lambda_1$ as otherwise equation (34) which is in this case

$$(\lambda_1-\lambda)(d-\lambda) = 0$$

would not have the double root λ_1. From (37) it now follows that

$$\frac{Y_2}{Y_1} = \frac{y_2}{y_1} + \frac{c}{\lambda_1}$$

---

* *Possible*, only, since r_1 and r_2 may be integral.

thus showing that the ratio y_2/y_1 is altered merely by the addition of a constant on circuit round x_0. This is also the case for the function

$$\frac{c}{2\pi\lambda_1 i}\log(x-x_0) = \frac{c}{2\pi\lambda_1 i}\log|x-x_0| + \frac{c}{2\pi\lambda_1}\arg(x-x_0)$$

Therefore, writing $c/2\pi\lambda_1 i = A$, we see that the function

$$\psi(x) = \frac{y_2}{y_1} - A\log(x-x_0)$$

is unaltered by a rotation about x_0 and consequently is *uniform* in the neighbourhood of this point. We therefore conclude that *when the fundamental equation (34) has a double root* $\lambda = \lambda_1$ *the differential equation (23) possesses two linearly independent integrals* $y_1(x)$ *and* $y_2(x)$ *where the first, as before, is represented in the neighbourhood of* x_0 *by the first equation in (36) while* $y_2(x)$ *is represented by*

(38) $$y_2(x) = y_1(x)[A\log(x-x_0) + \psi(x)]$$

where $\psi(x)$ *denotes an analytic function uniform about the point* x_0 *and* A *a constant* (*which may be zero*).

The three functions ϕ_1, ϕ_2, and ψ may be either *regular* at the point x_0, or may possess a *pole* there or even an *essential singularity*. In all cases these functions may be represented in the neighbourhood of $x = x_0$ by a Laurent's series of the form

$$\sum_{n=-\infty}^{\infty} a_n(x-x_0)^n$$

convergent in a suitable annulus with centre at the point x_0.

45. The case with no essential singularities

There is no doubt that the cases in which the functions ϕ_1, ϕ_2, and ψ are regular or possess poles at the point x_0 are more elementary than those in which essential singularities occur. It is therefore worth while considering whether these first two cases can be distinguished *a priori*, and, as we shall see, the necessary and sufficient conditions that no essential singularities exist turn out to be very simple and are satisfied in many important cases. We shall prove the following theorem:

FUCHS' THEOREM. *The necessary and sufficient conditions that the linear differential equation*

$$y'' + p_1(x) y' + p_2(x) y = 0$$

possesses a fundamental system of integrals $y_1(x)$, $y_2(x)$ representable in the neighbourhood of a singular point $x = x_0$ of the coefficients by formulæ of the form (36)–(38) where $\phi_1(x)$, $\phi_2(x)$ and $\psi(x)$ denote analytic uniform functions with at most a pole at the point x_0, are that the coefficient $p_1(x)$ has at most a pole of the FIRST *order and the coefficient $p_2(x)$ at most a pole of the* SECOND *order at the point x_0. If these conditions are satisfied we say that the point x_0 is a non-essential singularity of the equation.**

We begin by establishing the necessity of the Fuchs' conditions.

As the coefficients of an algebraic equation may be expressed in terms of the roots of the equation so also may the coefficients of a linear differential equation be easily expressed in terms of the integrals of a fundamental system.** In particular for the equation we are considering, by multiplying the two identities

(39) $\quad y_1'' + p_1(x) y_1' + p_2(x) y_1 = 0, \qquad y_2'' + p_1(x) y_2' + p_2(x) y_2 = 0$

respectively by $-y_2$ and y_1 and adding, we derive

$$y_2'' y_1 - y_2 y_1'' + p_1(x)(y_2' y_1 - y_2 y_1') = 0$$

whence, on dividing by the Wronskian $y_2' y_1 - y_2 y_1'$ of the system which is non-zero, we obtain

(40) $\quad -p_1(x) = \dfrac{y_2'' y_1 - y_2 y_1''}{y_2' y_1 - y_2 y_1'} = \dfrac{d}{dx} \log(y_2' y_1 - y_2 y_1')$

$$= \dfrac{d}{dx} \log \left[y_1^2 \dfrac{d}{dx} \left(\dfrac{y_2}{y_1} \right) \right]$$

This formula allows us to calculate $p_1(x)$ easily when y_1 and y_2 are known.***

---

* Some writers call this a *regular singular point*; others use the name a *Fuchsian singular point*.

** The formal similarities which exist in some respects between the theory of linear differential equations and the theory of algebraic equations have been stressed by E. PICARD: *Traité d'Analyse* (74), Vol. III, Chapter XVII.

*** From (40) can be easily derived the *Liouville formula* (see, for example, F. TRICOMI (79), Part II, p. 289) which determines, apart from a constant, the value of the Wronskian $W(x) = y_2' y_1 - y_2 y_1'$: for, from (40),

$$\dfrac{d}{dx} \log W(x) = -p_1(x)$$

whence

$$W(x) = c e^{-\int p_1(x) dx}$$

When $p_1(x)$ is thus determined $p_2(x)$ is obtained from one or other of the relations (39); for example, from the first equation in (39)

$$\text{(41)} \qquad -p_2(x) = \frac{y_1''}{y_1} + p_1(x)\frac{y_1'}{y_1}$$

We shall make use of the well-known result in the theory of analytic functions that *the logarithmic derivative of a function of the form*

$$\text{(42)} \qquad F(x) = (x-x_0)^\alpha G(x)$$

where α denotes ANY *constant and $G(x)$ a function regular at $x = x_0$ has at most a pole of first order at $x = x_0$.*

Supposing for greater generality that $G(x)$ has a zero of order $n \geq 0$ at the point x_0, we may write

$$G(x) = (x-x_0)^n G_1(x)$$

where $G_1(x)$ denotes a regular function, *non-zero* for $x = x_0$; thus

$$F(x) = (x-x_0)^{\alpha+n} G_1(x)$$

whence follow successively

$$\log F(x) = (\alpha+n)\log(x-x_0) + \log G_1(x)$$

$$\frac{F'(x)}{F(x)} = \frac{\alpha+n}{x-x_0} + \frac{G_1'(x)}{G_1(x)}$$

showing that $F'(x)/F(x)$ has a pole of first order (provided that $\alpha+n \neq 0$) at the point x_0.

On the other hand *the ratio $F''(x)/F(x)$ where $F(x)$ denotes a function of the form* (42) *has at most a pole of second order at $x = x_0$*; for

$$F'(x) = (\alpha+n)(x-x_0)^{\alpha+n-1} G_1(x) + (x-x_0)^{\alpha+n} G_1'(x)$$

$$F''(x) = (\alpha+n)(\alpha+n-1)(x-x_0)^{\alpha+n-2} G_1(x)$$
$$\qquad + 2(\alpha+n)(x-x_0)^{\alpha+n-1} G_1'(x) + (x-x_0)^{\alpha+n} G_1''(x)$$

whence

$$\frac{F''(x)}{F(x)} = \frac{(\alpha+n)(\alpha+n-1)}{(x-x_0)^2} + 2\frac{\alpha+n}{x-x_0}\frac{G_1'(x)}{G_1(x)} + \frac{G_1''(x)}{G_1(x)}$$

Further, it is evident that products, quotients, and derivatives of functions of the form (42) are also of this form.*

---

* Not, however, sums, unless the relative indices α differ by integers.

From these preliminary results we can deduce the *necessity* of Fuchs' conditions when the roots of the fundamental equation are distinct, as follows:

The functions on the right-hand sides of (36) are obviously of the form (42) although the functions $\phi_1(x)$ and $\phi_2(x)$ may have *poles* (of any order) at the point x_0;* the same may therefore be said of the functions

$$\frac{y_2}{y_1}, \quad \frac{d}{dx}\left(\frac{y_2}{y_1}\right) \quad \text{and} \quad y_1^2 \frac{d}{dx}\left(\frac{y_2}{y_1}\right)$$

But in view of (40), $-p_1(x)$ equals the logarithmic derivative of the last of these functions; therefore $p_1(x)$ has at most a pole of the first order at $x = x_0$.

The coefficient $p_2(x)$ can have at most a pole of the second order at x_0 since the first term on the right-hand side of (41) is of this form and so also is the second term, as the product of two functions with at most poles of the first order has at most a pole of the second order.

This deals with the case in which the fundamental equation has distinct roots.

If, instead, the fundamental equation has a double root in order to conclude that also in this case $p_1(x)$ has at most a pole of the first order we need only observe that it follows from (38) and (40) that

$$-p_1(x) = \frac{d}{dx}\log\left\{y_1^2 \frac{d}{dx}[A\log(x-x_0)+\psi(x)]\right\}$$
$$= \frac{d}{dx}\log\left\{y_1^2\left[\frac{A}{x-x_0}+\psi'(x)\right]\right\}$$

Thus $p_1(x)$ is again shown to be the logarithmic derivative of a function of the form (42). For $p_2(x)$ we need make no change in the previous proof as on the right-hand side of (41) there appears only the integral $y_1(x)$ which is still represented by the first of the two equations (36).

46. Integration in series of equations of Fuchs' type

We have still to show that the Fuchs' conditions are also *sufficient*; for this we shall use the method of Frobenius, i.e. the classical method of integration in series, introducing the modifications obviously suggested by the forms of

---

* If, for example, $\phi_1(x)$ possessed a pole of order m at the point x_0, then $\phi_1(x) = (x-x_0)^{-m}\Phi_1(x)$ where $\Phi_1(x)$ denotes a function regular at $x = x_0$; hence
$$y_1(x) = (x-x_0)^{r_1-m}\Phi_1(x)$$

the right-hand sides of (36) and (38). This has the advantage of illustrating a *practical* method of determination of the quantities r_1 and r_2 and the functions ϕ_1, ϕ_2, and ψ.*

For simplicity we suppose the singular point to be the origin (by change of variable if necessary), i.e. we suppose $x_0 = 0$, and the coefficients $p_1(x)$ and $p_2(x)$ to satisfy the Fuchs' conditions,

$$(43) \qquad p_1(x) = \frac{A(x)}{x}, \qquad p_2(x) = \frac{B(x)}{x^2}$$

where

$$(43') \qquad A(x) = \sum_{n=0}^{\infty} a_n x^n, \qquad B(x) = \sum_{n=0}^{\infty} b_n x^n \qquad (|x| < a)$$

(Note: we do not exclude the possibility that some of the leading coefficients in these two series may be zero.) We write the equation in the form

$$(44) \qquad x^2 y'' + A(x) x y' + B(x) y = 0$$

and try to satisfy it by the function

$$(45) \qquad y = x^r \sum_{n=0}^{\infty} c_n x^n$$

where r and the c_n are constants as yet undetermined apart from the one condition that $c_0 \neq 0$.**

From (45) follow

$$xy' = x^r \sum_{n=0}^{\infty} (r+n) c_n x^n, \qquad x^2 y'' = x^r \sum_{n=0}^{\infty} (r+n)(r+n-1) c_n x^n$$

whence, by substitution into (44) and by removing the common factor x^r, we derive

$$\sum_{n=0}^{\infty} (r+n)(r+n-1) c_n x^n + \sum_{n=0}^{\infty} (r+n) c_n x^n \cdot \sum_{n=0}^{\infty} a_n x^n + \sum_{n=0}^{\infty} c_n x^n \cdot \sum_{n=0}^{\infty} b_n x^n = 0$$

On application of the rule for multiplication of two power series, this gives

$$\sum_{n=0}^{\infty} \left\{ (r+n)(r+n-1) c_n + \sum_{m=0}^{n} [(r+m) a_{n-m} + b_{n-m}] c_m \right\} x^n = 0$$

---

* The previous formulæ (35') are of little practical use in the determination of the quantities r_1 and r_2 in view of the difficulty of finding the constants a, b, c, d which appear in the fundamental equation (34).

** This is in fact no restriction for if $c_0 = 0$, $c_1 \neq 0$, all that is required to satisfy the hypothesis is to change r into $r + 1$ and c_n into c_{n-1}.

In order that the series (45) should satisfy the given differential equation it is sufficient that the constants $r, c_0, c_1, c_2, \ldots$ satisfy the recurrence relations

$$(r+n)(r+n-1)c_n + \sum_{m=0}^{n} [(r+m)a_{n-m} + b_{n-m}]c_m = 0 \quad (n = 0, 1, 2, \ldots)$$

i.e.

$$[(r+n)(r+n-1) + (r+n)a_0 + b_0]c_n = -\sum_{m=0}^{n-1}[(r+m)c_{n-m} + b_{n-m}]c_m$$

which may be written in a more compact form by introducing the auxiliary functions (depending only on the coefficients a_ν and b_ν)

(46) $$\begin{cases} f_0(\xi) = \xi(\xi-1) + a_0\xi + b_0 \\ f_\nu(\xi) = a_\nu\xi + b_\nu \end{cases} \quad (\nu = 1, 2, 3, \ldots)$$

The preceding relations now take the form

(47) $$f_0(r+n)c_n = -\sum_{m=0}^{n-1} f_{n-m}(r+m)c_m$$

and give explicitly for the first few values of n,

(47') $$\begin{cases} f_0(r)c_0 = 0 \\ f_0(r+1)c_1 = -f_1(r)c_0 \\ f_0(r+2)c_2 = -f_2(r)c_0 - f_1(r+1)c_1 \\ f_0(r+3)c_3 = -f_3(r)c_0 - f_2(r+1)c_1 - f_1(r+2)c_2 \\ \cdots\cdots\cdots\cdots\cdots\cdots\cdots\cdots\cdots \end{cases}$$

The first of these relations determines r, since as $c_0 \neq 0$, $f_0(r) = 0$; i.e. r must be one of the roots of the equation

(48) $$r(r-1) + a_0 r + b_0 = 0$$

which may also be written as

(48') $$r^2 + [A(0) - 1]r + B(0) = 0$$

and is called the *indicial equation* (or *characteristic equation*).

The *indicial equation* which, in contrast to the fundamental equation (34), may be written down as soon as the differential equation is given is of basic importance in the study of Fuchsian equations. The indicial equation has two roots, r_1 and r_2, distinct or coincident, which we suppose named so that the real part of r_1 is not less than the real part of r_2, i.e. that

(49) $$\mathcal{R}(r_1) \geq \mathcal{R}(r_2)$$

and we put

(50) $$r_1 - r_2 = s$$

Thus $\mathscr{R}(s) \geq 0$. (The possibility that $s = 0$ is not excluded.)

We shall show that *corresponding to the root r_1 of the indicial equation there is in every case a solution of the given equation of the form* (45), i.e. whether s be zero or not, corresponding to r_1 there exists a series $\Sigma c_n x^n$ whose radius of convergence is non-zero.

We have
$$f_0(\xi) = (\xi - r_1)(\xi - r_2)$$
whence
$$f_0(r_1 + n) = n(n+s) \qquad (n = 1, 2, \ldots)$$

Thus $f_0(r_1 + n)$ can never be zero for

(51) $$\mathscr{R} f_0(r_1 + n) = n\mathscr{R}(n+s) \geq n^2 \geq 1$$

Hence the equations (47') successively determine the c_n for $n \geq 1$, when once the value of c_0^* ($\neq 0$) is fixed.

Further, if the two regular analytic functions $A(x)$ and $B(x)$ satisfy on a circle with centre x_0 and radius $\rho \leq a$ the inequalities

$$|A(x)| \leq M, \qquad |B(x)| \leq N$$

then (compare page 203)

$$|a_n| \leq \frac{M}{\rho^n}, \qquad |b_n| \leq \frac{N}{\rho^n}$$

and therefore

$$|f_\nu(\xi)| \leq |a_\nu||\xi| + |b_\nu| \leq \frac{M|\xi| + N}{\rho^\nu} \qquad (\nu \geq 1)$$

But on the other hand, it follows from (51) that

(52) $$|f_0(r_1 + n)| \geq n^2$$

whence from the recurrence formulæ (47) we deduce, for $r = r_1$, $n \geq 1$,

$$|c_n| \leq \frac{1}{n} \sum_{m=0}^{n-1} \left(\frac{M|r_1| + N}{n} + \frac{mM}{n} \right) \frac{|c_m|}{\rho^{n-m}} \qquad (n \geq 1)$$

The content of the bracket in the above sum is less than the fixed number

$$M|r_1| + N + M$$

---

*That the integral is determined only as far as a multiplicative constant is obvious from the differential equation.

which we denote by K; therefore

$$|c_n| < \frac{K}{n}\sum_{m=0}^{n-1}\frac{|c_m|}{\rho^{n-m}} \qquad (n=1,2,3,\ldots)$$

Now supposing that K is not less than 1,* which implies $1+K \leq 2K$, $1+K+K^2 \leq 3K^2$, etc., we derive successively

$$|c_1| < \frac{K}{\rho}|c_0|, \qquad |c_2| < \frac{K}{\rho^2}\left(\frac{1+K}{2}\right)|c_0| \leq \left(\frac{K}{\rho}\right)^2|c_0|$$

$$|c_3| < \frac{K}{\rho^3}\left(\frac{1+K+K^2}{3}\right)|c_0| \leq \left(\frac{K}{\rho}\right)^3|c_0|, \qquad \ldots$$

$$|c_n| < \left(\frac{K}{\rho}\right)^n|c_0|$$

But this shows that the series $\Sigma c_n x^n$ is majorized by the geometric series

$$|c_0|\sum_{n=0}^{\infty}\left(\frac{K|x|}{\rho}\right)^n$$

and consequently *the series (45) corresponding to $r = r_1$ has radius of convergence not less than ρ/K.*

We now consider the other root $r = r_2$ of the indicial equation, and *it is necessary to distinguish two cases according as the difference $s = r_1 - r_2$ is or is not an integral number (positive or zero)*; for

$$f_0(r_2+n) = n(n-s)$$

and it is clear that when s coincides with some positive integer ν the equations (47') corresponding to $r = r_2$ no longer determine c_ν and the successive coefficients, as $f_0(r_2+\nu) = 0$; while if $s = 0$, i.e. $r_1 = r_2$, the calculations may be carried out but give the same series as that found previously.

Excluding the exceptional cases arising when s is integral it is easy to show that the series (45) corresponding to $r = r_2$ behaves similarly to that obtained for $r = r_1$; it represents another integral of (44), linearly independent of the integral previously obtained.

For proof of this we require to repeat the previous calculations, substituting in place of (52) the corresponding relation obtained by considering the upper bound S of the enumerable set

$$\frac{1}{|1-s|}, \quad \frac{2}{|2-s|}, \quad \frac{3}{|3-s|}, \quad \ldots, \quad \frac{n}{|n-s|}, \quad \ldots,$$

---

* Otherwise put $K = 1$.

(which is certainly bounded as the denominators are all different from zero and the limit of the general term as $n \to \infty$ is 1) and observing that

$$|f_0(r_2+n)| = |n(n-s)| = \frac{n^2}{n/|n-s|} \geq \frac{n^2}{S}$$

We show that the two integrals denoted by y_1 and y_2, corresponding respectively to r_1 and r_2, are linearly independent by proving that the Wronskian of y_1 and y_2 cannot be identically zero; for

$$y_1 = c_0 x^{r_1} + \ldots, \quad y_1' = r_1 c_0 x^{r_1-1} + \ldots; \quad y_2 = c_0^* x^{r_2} + \ldots, \quad y_2' = r_2 c_0^* x^{r_2-1} + \ldots,$$

where the dots denote terms in higher powers of x than those shown explicitly and c_0^* the initial coefficient of the second series (45); therefore

$$\begin{vmatrix} y_1 & y_2 \\ y_1' & y_2' \end{vmatrix} = c_0 c_0^* (r_2 - r_1) x^{r_1+r_2-1} + \ldots$$

It remains to consider the case in which *the difference $s = r_1 - r_2$ of the roots of the indicial equation is an integer, positive or zero*, i.e. the case in which the two roots

$$\lambda_1 = e^{2\pi r_1 i}, \qquad \lambda_2 = e^{2\pi r_2 i}$$

of the fundamental equation (34) are coincident. We should expect that the second integral $y_2(x)$ of the given equation would be of the form (38); this suggests introducing as a new unknown the ratio $y_2(x)/y_1(x)$ where $y_1(x)$ is the solution previously determined.

It is convenient to use the formula of Liouville* which may be written as

$$y_1^2 \frac{d}{dx}\left(\frac{y_2}{y_1}\right) = c \exp\left[-\int p_1(x)\,dx\right]$$

Since

$$p_1(x) = \frac{A(x)}{x} = \frac{a_0}{x} + a_1 + a_2 x + a_3 x^2 + \ldots$$

then

$$\int p_1(x)\,dx = a_0 \log x + a_1 x + a_2 \frac{x^2}{2} + a_3 \frac{x^3}{3} + \ldots$$

whence, on substituting into the above formula, we derive

$$(53) \qquad \frac{d}{dx}\left(\frac{y_2}{y_1}\right) = \frac{c}{y_1^2} x^{-a_0} \exp\left(-\sum_{n=1}^{\infty} \frac{a_n}{n} x^n\right) = c x^{-a_0 - 2r_1} \frac{\exp\left(-\sum_{n=1}^{\infty} \frac{a_n}{n} x^n\right)}{\left(\sum_{n=0}^{\infty} c_n x^n\right)^2}$$

---

* Cf. the note on p. 219.

Here the quotient appearing on the extreme right above is an analytic function of x regular about the origin (since $c_0 \neq 0$), which may therefore be expanded in the form

(54) $$\frac{\exp\left(-\sum_{n=1}^{\infty}\frac{a_n}{n}x^n\right)}{\left(\sum_{n=0}^{\infty}c_n x^n\right)^2} = \alpha_0 + \alpha_1 x + \alpha_2 x^2 + \ldots$$

where the α's can be determined. Also, as the sum $r_1 + r_2$ of the roots of the indicial equation is in fact $-(a_0 - 1)$,

$$-a_0 - 2r_1 = r_1 + r_2 - 1 - 2r_1 = -(1+s)$$

Hence we may write (53) as

$$\frac{d}{dx}\left(\frac{y_2}{y_1}\right) = \frac{c}{x^{1+s}}(\alpha_0 + \alpha_1 x + \alpha_2 x^2 + \ldots)$$

or, since s is zero or a positive integer,

$$\frac{d}{dx}\left(\frac{y_2}{y_1}\right) = c\left(\frac{\alpha_0}{x^{1+s}} + \frac{\alpha_1}{x^s} + \frac{\alpha_2}{x^{s-1}} + \ldots + \frac{\alpha_s}{x} + \alpha_{s+1} + \alpha_{s+2}x + \ldots\right)$$

By integrating both sides we obtain the following result in which c' is a new arbitrary constant,

$$\frac{y_2}{y_1} = c\left(-\frac{\alpha_0}{sx^s} - \frac{\alpha_1}{(s-1)x^{s-1}} - \ldots - \frac{\alpha_{s-1}}{x} + \alpha_s \log x \right.$$
$$\left. + \alpha_{s+1}x + \frac{\alpha_{s+2}}{2}x^2 + \ldots + c'\right)$$

so that, in conformity with (38),

(55) $$y_2(x) = y_1(x)[A \log x + \psi(x)]$$

where $c\alpha_s = A$ (so that A vanishes if α_s does) and where $\psi(x)$ denotes the analytic function, with at most a pole of order s at the origin, defined by

(56) $$\psi(x) = c\left(-\frac{\alpha_0}{sx^s} - \frac{\alpha_1}{(s-1)x^{s-1}} - \ldots - \frac{\alpha_{s-1}}{x} + c' \right.$$
$$\left. + \alpha_{s+1}x + \frac{\alpha_{s+2}}{2}x^2 + \ldots\right)$$

in which c and c' are arbitrary constants (to which we may give the values 1 and 0, for example) and the coefficients α_s are obtained from (54).*

---

* That the two integrals y_1 and y_2 are linearly independent, i.e. that the ratio y_2/y_1 is not constant, (even in the case when $A = 0$) is easily seen since this can be the case if and only if all the α's vanish and this cannot happen as the left-hand side of (54) is not zero.

This concludes the proof of the fundamental theorem and in fact provides a practical method of integration by series of differential equations satisfying the Fuchs' conditions.

From a practical standpoint however it may be objected that in the case in which the indicial equation has equal roots, or roots differing by an integer, the actual determination of the function $\psi(x)$ by (56) is of little use in view of the difficulty of determining the values of the coefficients α_s from (54). It is usually preferable to find the second solution by substituting for y, into the equation (44), the expression

$$(57) \qquad A y_1(x) \log x + x^{r_2} \sum_{n=0}^{\infty} \gamma_n x^n$$

in which the γ_n are to be determined by equating to zero the part of the left-hand side not containing the factor $\log x$.

It should be observed that the lower limits found earlier for the radii of convergence of the series which represent y_1 and y_2 are of no practical importance as once we have established that these radii are in fact non-zero they can be immediately evaluated by the general theorem in § 43. This theorem established that these radii coincide with the minimum distance from the origin of the remaining singular points of the coefficients of the given equation.

Finally, it should be noted that the functions multiplying x^{r_1} and x^{r_2} in the representation of the integrals are *regular* at the origin (i.e. they have no poles there, as was previously admitted as a possibility). This is so also for the function $\psi(x)$ which appears in (55) when $s = 0$.

47. Totally Fuchsian equations. The hypergeometric equation

The theoretical considerations of the last two sections are of interest also from the practical point of view as most of the linear differential equations which have so far appeared in applied mathematics satisfy the Fuchs' conditions at their finite singular points; for example, the *Bessel equation* (see page 106 and following) may be written as

$$y'' + \frac{1}{x} y' + \left(1 - \frac{v^2}{x^2}\right) y = 0$$

§ 47] Totally Fuchsian equations: The hypergeometric equation

and obviously satisfies the Fuchs' conditions at the point $x = 0$; so also does the *Laguerre equation* (page 181 and following)

$$y'' + \frac{\alpha+1-x}{x} y' + \frac{\lambda}{x} y = 0$$

and likewise the *Legendre equation* with singular points at $x = \pm 1$, which may be written as

$$y'' - \frac{2x}{(1-x)(1+x)} y' + \frac{\lambda}{(1-x)(1+x)} y = 0$$

However the situation is in general different when we consider the *point at infinity*; for this purpose we identify the behaviour of the equation

$$y'' + p_1(x) y' + p_2(x) y = 0$$

about the point $x = \infty$ with that about the point $\xi = 0$ of the new equation obtained on making the change of variable $x = 1/\xi$. On making this transformation we have

$$\frac{dy}{dx} = -\xi^2 \frac{dy}{d\xi}, \qquad \frac{d^2 y}{dx^2} = \xi^4 \frac{d^2 y}{d\xi^2} + 2\xi^3 \frac{dy}{d\xi}$$

and the equation becomes

(58) $$\frac{d^2 y}{d\xi^2} + \left[\frac{2}{\xi} - \frac{1}{\xi^2} p_1\left(\frac{1}{\xi}\right) \right] \frac{dy}{d\xi} + \frac{1}{\xi^4} p_2\left(\frac{1}{\xi}\right) y = 0$$

That (58) may be Fuchsian about $\xi = 0$ we require that $p_1(1/\xi)$ and $p_2(1/\xi)$ possess *zeros at that point of orders at least equal to their suffices*, i.e. that $p_1(x)$ and $p_2(x)$ possess zeros at $x = \infty$ of such orders. This condition is actually satisfied for the Legendre equation but for neither the Bessel nor Laguerre equation. Nor, in fact, is the condition satisfied for elementary equations with constant coefficients—the explanation here lies in the fact that the transcendental functions defined by these equations possess essential singularities at infinity.

The question now arises whether it is possible to characterize (and classify) in some simple way linear differential equations of the second order with isolated singular points which, as does the Legendre equation, *satisfy the Fuchs' conditions at all their singular points including the point at infinity*. Equations of this type are usually said to be 'of Fuchsian type' but we prefer to call them 'totally Fuchsian equations'.

Such a characterization can be simply carried out since, as the coefficients $p_1(x)$ and $p_2(x)$ are analytic functions regular (even perhaps null) at infinity which possess only poles (respectively simple and at most double) for finite x, these coefficients can possess no essential singularities and hence must be *rational functions*. Therefore if $\alpha_1, \alpha_2, \ldots, \alpha_n$ are the points (necessarily finite in number) at which the poles of $p_1(x)$ and of $p_2(x)$ (or of both) occur, and if

(59) $$(x-\alpha_1)(x-\alpha_2)\ldots(x-\alpha_n) = P_n(x)$$

it is clear that both $p_1(x)$ and $p_2(x)$ may be represented as quotients of polynomials whose denominators are $P_n(x)$ and $P_n^2(x)$ respectively and whose numerators are some suitable polynomials $Q(x)$ and $R(x)$ respectively. As $p_1(x)$ and $p_2(x)$ possess at infinity zeros of orders at least equal to their suffices, the polynomials $Q(x)$ and $R(x)$ are at most of degrees $n-1$ and $2n-2$ respectively. Thus follows the theorem:

In order that (23) be a totally Fuchsian equation it is necessary and sufficient that its coefficients are rational functions of the form

(60) $$p_1(x) = \frac{Q_{n-1}(x)}{P_n(x)}, \quad p_2(x) = \frac{R_{2n-2}(x)}{P_n^2(x)}$$

where P_n, Q_{n-1} and R_{2n-2} are polynomials of degrees equal to their respective suffices ($n \geq 1$) and the zeros of the polynomial $P_n(x)$ are all simple (i.e. $P_n(x)$ may be represented in the form (59) where $\alpha_1, \alpha_2, \ldots, \alpha_n$ denote distinct points.)

The equation whose coefficients are given by (60) has in general $n+1$ singular points—the n points $\alpha_1, \alpha_2, \ldots, \alpha_n$ and the point at infinity.

The indicial equation relative to the singular point $\alpha_h (h = 1, 2, \ldots, n)$ may be written, in view of (48), as

(61) $$r(r-1) + \frac{Q_{n-1}(\alpha_h)}{P_n'(\alpha_h)} r + \frac{R_{2n-2}(\alpha_h)}{P_n'^2(\alpha_h)} = 0$$

since

$$(\alpha_h - \alpha_1)(\alpha_h - \alpha_2)\ldots(\alpha_h - \alpha_{h-1})(\alpha_h - \alpha_{h+1})\ldots(\alpha_h - \alpha_n) = P_n'(\alpha_h)$$

and the indicial equation relative to the point at infinity becomes, in view of (58),

(61') $$r(r-1) + (2+\alpha)r + \beta = 0$$

where

(62)
$$\begin{cases} \lim_{\xi \to 0} \frac{1}{\xi} p_1\left(\frac{1}{\xi}\right) = \lim_{x \to \infty} x p_1(x) = -\alpha \\ \lim_{\xi \to 0} \frac{1}{\xi^2} p_2\left(\frac{1}{\xi}\right) = \lim_{x \to \infty} x^2 p_2(x) = \beta \end{cases}$$

and $-\alpha$ and β denote the coefficients of x^{n-1} and x^{2n-2} in $Q_{n-1}(x)$ and $R_{2n-2}(x)$ respectively.

From the preceding formulæ follows the important result that *the sum of the roots of all the indicial equations of a totally Fuchsian equation with $n+1$ singular points equals $n-1$*.

If σ be this sum, by (61) and (61'),

$$\sigma = \sum_{h=1}^{n} \left[1 - \frac{Q_{n-1}(\alpha_h)}{P'_n(\alpha_h)}\right] - (\alpha+1) = n - 1 - \left[\sum_{h=1}^{n} \frac{Q_{n-1}(\alpha_h)}{P'_n(\alpha_h)} + \alpha\right]$$

But as the quotient

$$Q_{n-1}(\alpha_h)/P'_n(\alpha_h)$$

coincides with the residue at the pole α_h of the function $p_1(x)$ and α is the residue of the same function at infinity, the last square bracket is zero since the sum of all the residues of an analytic function with only isolated singularities equals *zero*;[*] therefore $\sigma = n-1$.

If however the equation while satisfying the Fuchs' conditions at each of the singular points $\alpha_1, \alpha_2, \ldots, \alpha_n$ does not satisfy them at infinity, then the sum σ' of the roots of the indicial equations relative to the points $\alpha_1, \alpha_2, \ldots, \alpha_n$ is, by the preceding argument,

$$\sigma' = n + \alpha$$

where α is the residue of the function $p_1(x)$ at infinity.

These considerations suggest the *classification* of totally Fuchsian equations according to the number of their singular points which, for $n \geq 1$, is at least two.

Totally Fuchsian equations with *two singular points* (i.e. $n = 1$) may be integrated elementarily and contribute nothing new. The most general equation of this type is of the form

$$y'' + \frac{h}{x-\alpha_1} y' + \frac{k}{(x-\alpha_1)^2} y = 0$$

---

[*] See, for example, F. TRICOMI (80), pp. 35, 75, 78.

where h and k denote arbitrary constants; this is essentially equation (27) for which we have already obtained the general integral, viz.

$$y = c_1(x-\alpha_1)^{r_1} + c_2(x-\alpha_2)^{r_2}$$

where c_1 and c_2 denote arbitrary constants and r_1 and r_2 are the roots of the quadratic equation (which we may now call the *indicial* equation)

$$r(r-1) + hr + k = 0$$

The simplest totally Fuchsian equations which can be expected to yield interesting results are those with *three singular points* (to which class Legendre's equation belongs). We shall consider such equations in some detail as they may be identified with one of the most important equations of analysis, *the hypergeometric equation*.

A totally Fuchsian equation with three singular points takes the forms

(63) $$y'' + \frac{h_1 x + h_2}{(x-\alpha_1)(x-\alpha_2)} y' + \frac{k_1 x^2 + k_2 x + k_3}{[(x-\alpha_1)(x-\alpha_2)]^2} y = 0$$

when the three points are α_1, α_2, and ∞, and

(64) $$y'' + \frac{2x^2 + h_1^* x + h_2^*}{(x-a)(x-b)(x-c)} y' + \frac{k_1^* x^2 + k_2^* x + k_3^*}{[(x-a)(x-b)(x-c)]^2} y = 0$$

when the singular points are any three distinct points a, b, c; for, in order that the point at infinity may be an ordinary point, $p_2(x)$ must have a zero of *fourth* order at infinity and $p_1(x)$ a zero of first order such that

$$\lim_{x \to \infty} x p_1(x) = 2 \quad \text{(see equation (58))}$$

The equations (63) and (64) show equally clearly that when once the positions of the singular points have been fixed, differential equations of this class depend on *five* parameters, viz. the undetermined coefficients of the first and second degree polynomials appearing in the numerators of $p_1(x)$ and $p_2(x)$.

Instead of these parameters we might use the three pairs of roots of the indicial equations which, as we have already seen, are related by the fact that their sum has the value $n-1$ (and for the case we are considering here, 1).

We shall now prove that *equation (64) is uniquely determined by the three singular points a, b, c together with the pairs of roots of the indicial equations relative to these singular points, (α, α'), (β, β'), (γ, γ') for which*

(65) $$\alpha + \alpha' + \beta + \beta' + \gamma + \gamma' = 1$$

Expressing the rational functions appearing in (64) in partial fractions we have

$$\frac{2x^2+h_1^*x+h_2^*}{(x-a)(x-b)(x-c)} = \frac{A}{x-a}+\frac{B}{x-b}+\frac{C}{x-c}$$

$$\frac{k_1^*x^2+k_2^*x+k_3^*}{(x-a)(x-b)(x-c)} = \frac{A'}{x-a}+\frac{B'}{x-b}+\frac{C'}{x-c}$$

where A, B, C and A', B', C' are suitable constants such that

(66) $$A+B+C = 2$$

(since the coefficient of x^2 in the numerator of the first rational function is 2). Thus (64) may be written as

(67) $$y'' + \left(\frac{A}{x-a}+\frac{B}{x-b}+\frac{C}{x-c}\right)y'$$
$$+ \left(\frac{A'}{x-a}+\frac{B'}{x-b}+\frac{C'}{x-c}\right)\frac{y}{(x-a)(x-b)(x-c)} = 0$$

which allows us to write in a particularly simple fashion the indicial equations relative to the three singular points a, b, c; for since

$$\lim_{x\to a}(x-a)p_1(x) = A, \quad \lim_{x\to a}(x-a)^2 p_2(x) = \frac{A'}{(a-b)(a-c)}$$

the indicial equation relative to $x = a$ is

$$r(r-1)+Ar+\frac{A'}{(a-b)(a-c)} = 0$$

which implies

$$\alpha+\alpha' = 1-A, \quad \alpha\alpha' = \frac{A'}{(a-b)(a-c)}$$

Similarly, on considering the other singular points b and c, we find

$$\beta+\beta' = 1-B, \quad \beta\beta' = \frac{B'}{(b-a)(b-c)}$$

$$\gamma+\gamma' = 1-C, \quad \gamma\gamma' = \frac{C'}{(c-a)(c-b)}$$

whence

(68) $$\begin{cases} A = 1-\alpha-\alpha', & B = 1-\beta-\beta', & C = 1-\gamma-\gamma' \\ A' = (a-b)(a-c)\alpha\alpha', & B' = (b-a)(b-c)\beta\beta', & C' = (c-a)(c-b)\gamma\gamma' \end{cases}$$

The relation (65) guarantees that (66) is satisfied; further, we can now write the differential equation in the *Papperitz form*

(69) $$y'' + \left(\frac{1-\alpha-\alpha'}{x-a} + \frac{1-\beta-\beta'}{x-b} + \frac{1-\gamma-\gamma'}{x-c}\right)y'$$
$$+ \left(\frac{(a-b)(a-c)\alpha\alpha'}{x-a} + \frac{(b-a)(b-c)\beta\beta'}{x-b}\right.$$
$$\left.+ \frac{(c-a)(c-b)\gamma\gamma'}{x-c}\right)\frac{y}{(x-a)(x-b)(x-c)} = 0$$

and the theorem is established. The six quantities α, α', β, β', γ, γ' are called the *characteristic exponents* of the equation.

Every function which satisfies (69) is called a *Riemann P-function* and is denoted by the symbol

(70) $$P\left\{\begin{matrix} a & b & c & \\ \alpha & \beta & \gamma & x \\ \alpha' & \beta' & \gamma' & \end{matrix}\right\}$$

This notation sets out clearly the nine parameters—which in view of (65) reduce to eight—on which, in addition to x, such a function depends.*

This large number of parameters can be reduced by employing two important properties of (69) derived by making a change of independent variable, viz.

$$x_1 = \frac{Ax+B}{Cx+D} \qquad (AD-BC \neq 0)$$

and by making a change of dependent variable, viz.

(71) $$y_1 = (x-a)^\lambda (x-b)^\mu (x-c)^\nu y \qquad \lambda+\mu+\nu = 0$$

These transformations establish the following formulæ**

---

* In view of the symmetry of the Papperitz equation in a, b, c and the pairs α, α'; β, β'; γ, γ' of characteristic exponents, it is clear that in the symbol (70) the three columns may be permuted; also in each column the second and third elements may be interchanged.

** (72) is easily established by proving the formula first for the three *elementary substitutions* $x_1 = x + h$, $x_1 = kx$, $x_1 = 1/x$; any linear substitution is made up of a product of all or some of these elementary substitutions, and the result follows. (73) can either be verified directly (entirely elementary calculation but not simple), or by remarking that the function y_1 obviously satisfies a linear differential equation of second order whose only (Fuchsian) *finite* singular points are a, b, c and further (and this demands some calculation) that under the condition $\lambda + \mu + \nu = 0$ this equation is regular at infinity.

It should be noted that in the formulæ (72) and (73) the equality sign is used in a slightly different sense from the usual—for here the symbol $P\{\ldots\}$ does not denote a particular solution of the equation (69) but *any* solution.

(72)
$$P\begin{Bmatrix} a & b & c \\ \alpha & \beta & \gamma & x \\ \alpha' & \beta' & \gamma' \end{Bmatrix} = P\begin{Bmatrix} a_1 & b_1 & c_1 \\ \alpha & \beta & \gamma & x_1 \\ \alpha' & \beta' & \gamma' \end{Bmatrix}$$

and

(73)
$$P\begin{Bmatrix} a & b & c \\ \alpha & \beta & \gamma & x \\ \alpha' & \beta' & \gamma' \end{Bmatrix}$$
$$= (x-a)^{-\lambda}(x-b)^{-\mu}(x-c)^{-\nu} P\begin{Bmatrix} a & b & c \\ \alpha+\lambda & \beta+\mu & \gamma+\nu & x \\ \alpha'+\lambda & \beta'+\mu & \gamma'+\nu \end{Bmatrix}, \quad (\lambda+\mu+\nu = 0)$$

where a_1, b_1, c_1 denote the points which correspond to a, b, c in the linear substitution above.

Consequently if the substitution be chosen so that a, b, c transform into the three points 0, 1, ∞*, and if, in addition, we use (73) with $\lambda = -\alpha$, $\mu = -\beta$, $\nu = \alpha+\beta$, we obtain

(74)
$$P\begin{Bmatrix} a & b & c \\ \alpha & \beta & \gamma & x \\ \alpha' & \beta' & \gamma' \end{Bmatrix}$$
$$= \left(\frac{x-a}{x-c}\right)^{\alpha}\left(\frac{x-b}{x-c}\right)^{\beta} P\begin{Bmatrix} 0 & 1 & \infty \\ 0 & 0 & \alpha+\beta+\gamma & x_1 \\ \alpha'-\alpha & \beta'-\beta & \alpha+\beta+\gamma' \end{Bmatrix}$$

where

(75) $\quad (\alpha'-\alpha)+(\beta'-\beta)+(\alpha+\beta+\gamma)+(\alpha+\beta+\gamma') = 1$

The Riemann P-function is therefore reduced to a form in which only three essential parameters appear.

To introduce the usual notation, we write

(76) $\quad \alpha+\beta+\gamma = a, \quad \alpha+\beta+\gamma' = b, \quad 1+\alpha-\alpha' = c$

so that the P-function takes the form

(77)
$$P\begin{Bmatrix} 0 & 1 & \infty \\ 0 & 0 & a & x \\ 1-c & c-a-b & b \end{Bmatrix}$$

* We employ the substitution $x_1 = \dfrac{b-c}{a-b}\dfrac{a-x}{x-c}$

in which a, b, c are new parameters unconnected with the quantities previously denoted by these same letters.

When the three singular points are initially $0, 1, \infty$, equation (74) takes the form

$$(74') \quad P\left\{\begin{matrix} 0 & 1 & \infty & \\ \alpha & \beta & \gamma & x \\ \alpha' & \beta' & \gamma' & \end{matrix}\right\} = x^\alpha(1-x)^\beta P\left\{\begin{matrix} 0 & 1 & \infty & \\ 0 & 0 & a & x \\ 1-c & c-a-b & b & \end{matrix}\right\}$$

where a, b, c are determined by (76); this is easily verified on putting, in (74), $a = 0, b = 1, c = \omega$ and making ω tend to infinity after having multiplied the right-hand side (as is permissible—see the note on p. 234) by the constant factor

$$(-1)^\beta \omega^{\alpha+\beta}$$

Similarly, putting $a = 0, b = 1, c = \omega$ and making ω tend to infinity, the Papperitz equation satisfied by the function (77) in which

$$\alpha = \beta = 0, \qquad \gamma = a, \qquad \alpha' = 1-c, \qquad \beta' = c-a-b, \qquad \gamma' = b$$

takes the simple form

$$(78) \qquad x(1-x)y'' + [c-(a+b+1)x]y' - aby = 0$$

This is the well-known *hypergeometric equation* with which we shall deal only briefly despite its importance.*

We here exclude the three cases of the hypergeometric equation

$$(79) \qquad c \text{ integral}, \quad c-a-b \text{ integral}, \quad a-b \text{ integral}$$

in whose solutions logarithmic terms may arise. The natures of the pairs of fundamental integrals about the three singular points $0, 1, \infty$ are determined by the previous work and are as follows: about the origin the two integrals are of the forms

$$(80) \qquad y_1 = P_1(x), \qquad y_2 = x^{1-c} P_2(x)$$

About the points $x = 1$ and $x = \infty$ the integrals are of the forms

$$(80') \qquad y_3 = P_3(1-x), \qquad y_4 = (1-x)^{c-a-b} P_4(1-x)$$

and

$$(80'') \qquad y_5 = \left(\frac{1}{x}\right)^a P_5\left(\frac{1}{x}\right), \qquad y_6 = \left(\frac{1}{x}\right)^b P_6\left(\frac{1}{x}\right)$$

---

* For information on the *hypergeometric function* see F. KLEIN-O. HAUPT: *Vorlesungen über die hypergeometrische Funktion* (Berlin, Springer, 1933); also some chapters in A. R. FORSYTH (16), E. PICARD (74), E. T. WHITTAKER-G. N. WATSON (85), and G. SANSONE: *Lezioni sulla teoria di una variabile complessa*, 3° ed. (Padua, Cedam, 1950).

where $P_1, P_2, \ldots, P_6$ denote *power series* (of *positive integral powers*) whose circles of convergence—which extend as far as the nearest remaining singular point—are defined for each pair of integrals respectively by the inequalities

(81) $$|x|<1, \quad |1-x|<1, \quad |1/x|<1$$

Hence all the power series have radius of convergence equal to *one*.

The first of these power series is easily determined—supposing that c is not zero nor a negative integer*—by the usual method of substituting $y = \Sigma c_n x^n$ into (78); this leads to the recurrence relation

$$c_{n+1} = \frac{(a+n)(b+n)}{(1+n)(c+n)} c_n$$

and, supposing that $c_0 = 1$, we derive the corresponding solution as

(82) $$F(a,b;c;x) = 1 + \sum_{n=1}^{\infty} \frac{a(a+1)\ldots(a+n-1)\cdot b(b+1)\ldots(b+n-1)}{1\cdot 2 \ldots n \cdot c(c+1)\ldots(c+n-1)} x^n$$

This is the celebrated *hypergeometric series* first studied by Euler and subsequently by Gauss, Riemann, and others. In the particular case $a = 1$, $b = c$, the hypergeometric series reduces to the geometric series Σx^n—whence the nomenclature.

In the terminology of today the symbol $F(a, b; c; x)$ usually denotes the analytic function (in general multivalent) which about the origin is represented by the series (82), i.e. the function which is the analytic continuation throughout the entire complex plane of the series (82). The hypergeometric function which is symmetrical in a and b includes as special cases many important functions, some being elementary functions.** For example, in addition to the special case already mentioned in which $a = 1$, $b = c$, on writing $b = c$ we obtain the more general result that

$$F(a,b;b;x) = 1 + \sum_{n=1}^{\infty} \binom{-a}{n}(-x)^n = (1-x)^{-a}$$

Also, as is easily verified,

$$F(1,1;2;x) = -\frac{1}{x}\log(1-x)$$

and

$$F(\tfrac{1}{2},\tfrac{1}{2};1;k^2) = \frac{2}{\pi} K$$

where K denotes the elliptic integral which appeared in Chapter I (p. 23).

---

* Otherwise we would discuss the series P_2 since now $(1 - c)$ is the root of the indicial equation with larger real part.

** The same may be said of the *confluent* hypergeometric function Φ, which is a limiting case of the function F as will be seen later, in § 52.

Making use of the function $F(a, b; c; x)$ we can easily find the other five series $P_2, \ldots, P_6$ which appear in (80); here we shall evaluate only $P_2(x)$ and shall show that

$$P_2(x) = F(a-c+1, b-c+1; 2-c; x)$$

This implies that *when c is not integral, the general integral of the hypergeometric equation* (78) *is*

(83) $\qquad y = C_1 F(a, b; c; x) + C_2 x^{1-c} F(a-c+1, b-c+1; 2-c; x)$

where C_1 and C_2 denote arbitrary constants.

To prove the result we note that on the left-hand side of (74') we can interchange α and α', or β and β', or γ and γ'; also, in (76) the interchange of γ and γ' only permutes a and b. This latter result is of course essentially the same as observing that a and b appear symmetrically in the hypergeometric function $F(a, b; c; x)$ and in the differential equation (78); but the other two interchanges lead to interesting results. On interchanging α and α' we obtain

$$x^\alpha (1-x)^\beta F \begin{Bmatrix} 0 & 1 & \infty & \\ 0 & 0 & a & x \\ 1-c & c-a-b & b & \end{Bmatrix}$$

$$= x^{\alpha'}(1-x)^\beta P \begin{Bmatrix} 0 & 1 & \infty & \\ 0 & 0 & a' & x \\ 1-c' & c'-a'-b' & b' & \end{Bmatrix}$$

in which

$$a' = \alpha' + \beta + \gamma = a-c+1, \qquad b' = \alpha' + \beta + \gamma' = b-c+1$$
$$c' = 1 + \alpha' - \alpha = 2-c$$

so that

$$P \begin{Bmatrix} 0 & 1 & \infty & \\ 0 & 0 & a & x \\ 1-c & c-a-b & b & \end{Bmatrix} = x^{1-c} P \begin{Bmatrix} 0 & 1 & \infty & \\ 0 & 0 & a' & x \\ 1-c' & c'-a'-b' & b' & \end{Bmatrix}$$

This shows that contained among the functions represented by the symbol on the left-hand side above, i.e. contained among the particular integrals of (78), is the function

$$G(x) = x^{1-c} F(a-c+1, b-c+1; 2-c; x)$$

which *when c is not integral* is linearly independent of $F(a, b; c; x)$, for evidently, as $x \to 0$,

$$G(x) = x^{1-c}[1 + O(x)]$$

The general integral of (78) is therefore of the form (83).

The interchange of β and β' leads to the similar result, that

$$P\left\{\begin{matrix} 0 & 1 & \infty & \\ 0 & 0 & a & x \\ 1-c & c-a-b & b & \end{matrix}\right\} = (1-x)^{c-a-b} P\left\{\begin{matrix} 0 & 1 & \infty & \\ 0 & 0 & a'' & x \\ 1-c'' & c''-a''-b'' & b'' & \end{matrix}\right\}$$

in which

$$a'' = \alpha + \beta' + \gamma = c - b, \qquad b'' = \alpha + \beta' + \gamma' = c - a$$
$$c'' = 1 + \alpha - \alpha' = c$$

Consequently there exist two constants A and B such that

$$(1-x)^{c-a-b} F(c-b, c-a; c; x)$$
$$\equiv AF(a, b; c; x) + Bx^{1-c} F(a-c+1, b-c+1; 2-c; x)$$

In this identity the left-hand side tends to 1 as $x \to 0$ while the right-hand side is of the form

$$A[1 + O(x)] + Bx^{1-c}[1 + O(x)]$$

which implies $A = 1$, $B = 0$; the important result follows,

(84) $\qquad F(a, b; c; x) = (1-x)^{c-a-b} F(c-a, c-b; c; x)$

By continuity arguments it follows that this formula remains valid even if c is integral. Hence (84) states that every function F may be represented in two ways as a power series in the same variable.

By using (74') and (83) it is easy to derive an explicit formula for the general Riemann P-function with singular points 0, 1, ∞, in terms of F-functions; for

$$P\left\{\begin{matrix} 0 & 1 & \infty & \\ \alpha & \beta & \gamma & x \\ \alpha' & \beta' & \gamma' & \end{matrix}\right\} = (1-x)^{\beta}[C_1 x^{\alpha} F(\alpha + \beta + \gamma, \alpha + \beta + \gamma'; 1 + \alpha - \alpha'; x)$$
$$+ C_2 x^{\alpha'} F(\alpha' + \beta + \gamma, \alpha' + \beta + \gamma'; 1 + \alpha' - \alpha; x)]$$

provided that the difference $\alpha' - \alpha$ of the two characteristic exponents relative to the singular point $x = 0$ is not integral.

In conclusion, we mention that the study of totally Fuchsian equations with $n+1 > 3$ singular points leads to some possible generalizations of hypergeometric functions—*generalized hypergeometric functions*.* Here however arises a difficulty not encountered in the case in which $n = 2$, viz. the fact that while the number of parameters on which, in addition to the $n+1$ singular points, the differential equation depends (cf. page 230) is $3n-1$, the number of independent characteristic exponents is only $2n+1$. The equation therefore contains $n-2$ so-called *supplementary parameters*.

48. Preliminary remarks on points of essential singularity

The discussion in the preceding section and the particular example of the hypergeometric equation clearly indicate the behaviour of the integrals of a linear differential equation in the neighbourhood of a *non-essential singularity* of the equation, i.e. a singular point at which the Fuchs' conditions are satisfied. But what happens if the point is an *essential singular point*, i.e. a point at which the Fuchs' conditions are *not* satisfied?

What we can so far say on this question is very little (meantime disregarding the methods of §29 which may be applicable). In general, the only previous results which are of help here are the formulæ (36)–(38); these show that the integrals may be multivalent functions, but otherwise say little about their actual behaviour (i.e. whether they tend to a limit, finite or infinite, or not, etc.) as the three functions $\phi_1(x)$, $\phi_2(x)$, $\psi(x)$ may now (and at least one of them actually *does*) possess essential singularities at the point under consideration.

It is well worth while to continue the previous work and try to derive series to represent in some way the pairs of linearly independent integrals of a linear differential equation about a point of essential singularity. Here we must expect that these series (if they can be derived) will be in general *divergent* and will represent the integrals of the equation only *asymptotically*, in a sense to be defined later. However these series are of importance even from the point of view of obtaining numerical results, as we shall see by an example which will again be Bessel's equation.

We shall suppose that the singular point to be considered is the *point at infinity* (by change of variable, if necessary, the transformation being that

---

* See for example, P. APPELL-J. KAMPÉ DE FÉRIET: *Fonctions hypergéométriques et hypersphériques. Polynomes d'Hermite* (Paris, Gauthier-Villars, 1926); W. N. BAILEY: *Generalized Hypergeometric Series*, Cambridge Tracts No. 32 (1935). Other possible generalizations of hypergeometric functions lead to consideration of functions of more than one independent variable, for example, *Appell functions*.
See also Bateman Project (A. ERDÉLYI, W. MAGNUS; F. OBERHETTINGER; F. TRICOMI,) *Higher Transcendental Functions*, I–III, (New York, McGraw-Hill, 1953-55).

used in reverse on page 229) and that *the coefficients $p_1(x)$ and $p_2(x)$ in the usual equation*

$$y'' + p_1(x)y' + p_2(x)y = 0$$

are analytic functions regular at infinity, i.e. may be represented for sufficiently large $|x|$ by the series

(85) $\qquad p_1(x) = a_0 + \dfrac{a_1}{x} + \dfrac{a_2}{x^2} + \ldots, \qquad p_2(x) = b_0 + \dfrac{b_1}{x} + \dfrac{b_2}{x^2} + \ldots$

This implies that, in general, $x = \infty$ is a point of *essential singularity* of the equation, since in order to make this a non-essential singular point the conditions of page 229 demand here that

(86) $\qquad\qquad\qquad a_0 = 0, \qquad b_0 = b_1 = 0$

while if $x = \infty$ is an ordinary point of the equation there are the stronger restrictions (see page 232) that

(87) $\qquad\qquad a_0 = 0, a_1 = 2, \qquad b_0 = b_1 = b_2 = b_3 = 0$

It is not difficult to think of conditions more general than those we are imposing on the equation;* for example, we might suppose that the functions $p_1(x)$ and $p_2(x)$ have poles of a certain order at the point $x = \infty$. But although the method that we shall use may be adapted to deal with those more general cases** we shall adhere to the above assumptions that the coefficients $p_1(x)$ and $p_2(x)$ may be represented by series of the type (85); for these assumptions simplify the working quite considerably and in fact cover all cases of importance.

It is essential to define in what sense the series to be derived will represent asymptotically (as $x \to \infty$) the integrals of the given equation. We give the classic definition due to Poincaré:

We say that *the series,* CONVERGENT OR NOT,

$$c_0 + \frac{c_1}{x} + \frac{c_2}{x^2} + \ldots$$

asymptotically represents (as $x \to \infty$) *the function $f(x)$ in the sense of Poincaré* and we write

(88) $\qquad\qquad f(x) \sim c_0 + \dfrac{c_1}{x} + \dfrac{c_2}{x^2} + \ldots + \dfrac{c_n}{x^n} + \ldots$

---

* The conditions we are using imply that at a finite singular point, $p_1(x)$ has at most a pole of *second* order and $p_2(x)$ one of *fourth* order.

** The method may be adapted to deal even with the case in which the series (85) represent the coefficients *asymptotically*, the meaning of which phrase will be defined immediately.

if for any positive integral n

(88′) $$f(x) = c_0 + \frac{c_1}{x} + \frac{c_2}{x^2} + \ldots + \frac{c_n}{x^n}[1 + \varepsilon_n(x)]$$

where $\varepsilon_n(x)$ denotes a function which tends to zero as $x \to \infty$, i.e. if

$$f(x) = c_0 + \frac{c_1}{x} + \frac{c_2}{x^2} + \ldots + \frac{c_n}{x^n} + o(x^{-n})$$

or, written otherwise,

(88″) $$f(x) = c_0 + \frac{c_1}{x} + \frac{c_2}{x^2} + \ldots + \frac{c_n}{x^n} + O(x^{-n-1})$$

We cannot deal here with the general properties of such asymptotic series and for these we refer the reader to the original work of Poincaré,* and to that of Knopp. For our present purposes we recall that these series may be added, multiplied, and even divided by similar series exactly as may be done with ordinary power series; further, they may be integrated term by term (but not, in general, differentiated term by term) since as we have already noted (compare pages 175, 177) that asymptotic equalities containing the symbol O may in general be integrated but not differentiated. Further, the asymptotic representation of a function $f(x)$ is *unique*, in the sense that if

$$f(x) \sim c_0 + \frac{c_1}{x} + \frac{c_2}{x^2} + \ldots, \quad f(x) \sim c_0' + \frac{c_1'}{x} + \frac{c_2'}{x^2} + \ldots$$

then necessarily $c_0 = c_0'$, $c_1 = c_1'$, $c_2 = c_2'$, ... Uniqueness, however, does not hold in the reverse direction, in the sense that a series of the form (88) may well represent asymptotically two different functions—for example, the functions $f(x)$ and $f(x) + e^{-x}$ would have the same asymptotic representation provided that x tends to infinity in such a way that $\lim \mathscr{R}x = +\infty$.

We now proceed to show how by a transformation of the type

(89) $$y = e^{\alpha x} x^\rho z = e^{\alpha x + \rho \log x} z$$

where α and ρ denote two constants to be suitably chosen later, the given equation may *in general* be simplified, in the sense that the coefficients of the transformed equation

$$\frac{d^2 z}{dx^2} + p_1^*(x) \frac{dz}{dx} + p_2^*(x) z = 0$$

satisfy at least the one Fuchs' condition that $p_2^*(x)$ has a zero of second order at infinity.

---

* H. POINCARÉ, 'Sur les intégrales irrégulières des équations linéaires', *Acta Math.*, **8**, 295–344 (1886); K. KNOPP (67), § 65.

Preliminary remarks on points of essential singularity

As it is easily verified that

(90)
$$\begin{cases} p_1^*(x) = p_1(x) + 2\left(\alpha + \dfrac{\rho}{x}\right) = a_0 + 2\alpha + \dfrac{a_1 + 2\rho}{x} + \sum_{n=2}^{\infty} a_n x^{-n} \\[2mm] p_2^*(x) = p_2(x) + \left(\alpha + \dfrac{\rho}{x}\right) p_1(x) + \alpha^2 + \dfrac{2\alpha\rho}{x} + \dfrac{\rho(\rho-1)}{x^2} \\[2mm] \qquad = (\alpha^2 + a_0\alpha + b_0) + \dfrac{(a_0 + 2\alpha)\rho + a_1\alpha + b_1}{x} + \dfrac{\rho(\rho-1)}{x^2} \\[2mm] \qquad + \sum_{n=2}^{\infty} \dfrac{a_n\alpha + a_{n-1}\rho + b_n}{x^n} \end{cases}$$

thus to fulfil the condition stated above we must choose α to be one of the two roots α_1, α_2 of the quadratic equation

(91) $$\alpha^2 + a_0\alpha + b_0 = 0$$

which equation we shall call the *characteristic equation** (as in the theory of differential equations with constant coefficients). Now *if the roots of the characteristic equations are distinct*, i.e. if

(92) $$2\alpha_h + a_0 \neq 0 \qquad (h = 1, 2)$$

the coefficient of $1/x$ in the above expression for $p_2^*(x)$ can be made to vanish by giving ρ the corresponding value

(93) $$\rho_h = -\frac{a_1\alpha_h + b_1}{2\alpha_h + a_0} \qquad (h = 1, 2)$$

As regards the first term in the expression for $p_1^*(x)$, this cannot be zero in view of (92), the condition that the roots of (91) are distinct. However we may reduce this term to -1 by making a final change of independent variable

(94) $$x = -\frac{1}{2\alpha_h + a_0} x'$$

This gives the final reduction of the equation to the form

(95) $$\frac{d^2z}{dx^2} - \left[1 - \frac{1}{x} f(x)\right] \frac{dz}{dx} + \frac{1}{x^2} g(x) z = 0$$

---

* Note that (91) in fact is identical with the characteristic equation of the *limit equation*
$$y'' + a_0 y' + b_0 y = 0$$
obtained by taking the limits of the coefficients of the original equation as $x \to \infty$.

(where the dash on x has been removed) in which $f(x)$ and $g(x)$ are analytic functions regular about the point at infinity and therefore representable for $|x|$ greater than a certain a, by series of the form

(96) $\qquad f(x) = A_0 + \dfrac{A_1}{x} + \dfrac{A_2}{x^2} + \dots, \qquad g(x) = B_0 + \dfrac{B_1}{x} + \dfrac{B_2}{x^2} + \dots$

whose coefficients may be expressed in terms of the preceding coefficients a_n and b_n.

We now apply the method of transformation into an integral equation, as in § 29 (second method), noting that in the terminology of § 29,

$$p_1(x) = -1, \quad p_2(x) = 0, \quad A(x) = 0, \quad B(x) = -x^{-1}f(x), \quad C(x) = -x^{-2}g(x)$$

and consequently

$$F_1(x) = 1, \qquad F_2(x) = e^x, \qquad W(x) = e^x$$

$$L(x, \xi) = e^x - e^\xi, \qquad M(x) = -x^{-2}g(x)e^{-x}, \qquad N(x) = -x^{-1}f(x)e^{-x}$$

$$L(x,\xi)M(\xi) - \frac{\partial}{\partial \xi}[L(x,\xi)N(\xi)] = -(e^x - e^\xi)\xi^{-2}g(\xi)e^{-\xi} + \frac{\partial}{\partial \xi}[(e^x - e^\xi)\xi^{-1}f(\xi)e^{-\xi}]$$

$$= \xi^{-2}[f(\xi) + g(\xi)] - \xi^{-1}f'(\xi) - e^{x-\xi}\{\xi^{-2}[f(\xi) + g(\xi)] + \xi^{-1}[f(\xi) - f'(\xi)]\}$$

Now, in addition, assuming $x_0 = \infty$ and referring to a solution $z(x)$ for which $z(x_0)$ remains bounded (and therefore $\gamma_2 = 0$), and putting for simplicity $\gamma_1 = 1$,[*] we derive from the integral equation (25) of § 29

(97) $\qquad z(x) - \displaystyle\int_x^\infty [\xi^{-2}F(\xi) + e^{x-\xi}\xi^{-1}G(\xi)] z(\xi)\, d\xi = 1$

where for brevity we write

(98) $\qquad \begin{cases} \xi f'(\xi) - f(\xi) - g(\xi) = F(\xi) \\ f(\xi) - f'(\xi) + \xi^{-1}[f(\xi) + g(\xi)] = f(\xi) - \xi^{-1}F(\xi) = G(\xi) \end{cases}$

It should be noted that the term $\xi f'(\xi)$ is a regular function (in fact is zero) for $\xi = \infty$, since

$$f'(\xi) = -A_1 \xi^{-2} - 2A_2 \xi^{-3} - \dots$$

To (97) however, the general theory of Volterra integral equations as outlined in § 29 cannot immediately be applied as here we are dealing with a 'singular' equation, the interval of integration now being infinite.

---

[*] A different choice for this constant would imply multiplication of the solution considered by a constant factor.

49. An application of the method of successive approximations

To simplify the formal work as much as possible it is convenient to introduce a special symbol T to denote the *linear functional transformation*[*] represented by the integral which appears in (97), i.e. the operation which acting on a function (here an analytic function) $\Phi(\xi)$ transforms it into $\phi(x)$; thus

$$(99) \quad \phi(x) = T[\Phi(\xi)] \equiv \int_x^\infty [\xi^{-2} F(\xi) + e^{x-\xi} \xi^{-1} G(\xi)] \Phi(\xi) \, d\xi$$

This enables us to write the integral equation (97) in the symbolic form

$$(100) \quad z(x) = 1 + T[z(\xi)]$$

and on applying to this equation the classical method of successive approximations we put

$$(101) \quad z_0(x) = 1, \quad z_1(x) = 1 + T[1], \quad z_2(x) = 1 + T[z_1(\xi)], \quad \ldots$$

and in general

$$(102) \quad z_{n+1}(x) = 1 + T[z_n(\xi)] \qquad (n = 0, 1, 2, \ldots)$$

which implies

$$(103) \quad z_{n+1}(x) - z_n(x) = T[z_n(\xi) - z_{n-1}(\xi)] \qquad (n = 1, 2, 3, \ldots)$$

In discussing the cardinal question of the convergence or otherwise of the sequence (101) it is helpful to start with a proof of the following general theorem on the functional transformation T:

The transformation $\phi(x) = T[\Phi(\xi)]$ of an analytic function $\Phi(\xi)$ regular in the neighbourhood $|\xi| \geq R$ of the point at infinity and there satisfying the condition

$$(104) \quad |\Phi(\xi)| \leq N |\xi|^{-n}$$

where N and n are two non-negative constants and n is integral, is regular in

---

[*] A functional transformation T is called *linear* when (as is obviously true in this case) $T(\Phi_1 + \Phi_2) = T(\Phi_1) + T(\Phi_2)$ and $T(C\Phi) = CT(\Phi)$ where C is any constant.

a neighbourhood of the point at infinity in the complex x-plane defined by the inequalities

$$|x| \geq R^* \geq R, \qquad -\pi + \eta \leq \arg x \leq \pi - \eta$$

where $0 < \eta < \pi/2$, and satisfies the condition

(105) $$|\phi(x)| \leq KN|x|^{-n-1}$$

where K is a positive constant independent of the special function Φ considered.

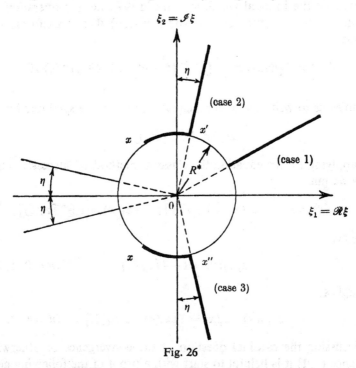

Fig. 26

The path of integration for the integral in the transformation T (which is partially arbitrary and will be denoted by Γ) starts at the point $\xi = x$ and continues to infinity in the complex ξ-plane; it must never cross into the circle $|\xi| \leq |x|$ and must be such that on it the real part of ξ tends to ∞. We define Γ precisely as follows:

(1) If $-\pi/2 + \eta < \arg x < \pi/2 - \eta$, we take Γ as the half-line from x to infinity along the radius from 0 to x.

(2) If $\pi/2 - \eta \leq \arg x \leq \pi - \eta$, then Γ is made up (see figure 26) of the arc of the circle $|\xi| = |x|$ between x and x', where $\arg x' = \pi/2 - \eta$, and of the half-line from x' to infinity along the radius $0x'$.

(3) If $-\pi + \eta \leq \arg x \leq -\pi/2 + \eta$, then Γ is made up as in (2) but here with x' replaced by the point x'' such that $|x''| = |x|$, $\arg x'' = -\pi/2 + \eta$.

We suppose that the analytic functions $F(\xi)$ and $G(\xi)$, regular at infinity, which appear in the transformation T, satisfy the conditions $|F(\xi)| \leq M$, $|G(\xi)| \leq M$ for $|\xi| \geq R'$, so that for $|\xi| \geq R^*$, where R^* is the greater of the two numbers R and R',

(106) $$|F(\xi)| \leq M, \quad |G(\xi)| \leq M, \quad |\Phi(\xi)| \leq N|\xi|^{-n}$$

It follows from this and from the fact that on Γ

$$|\xi|^{-n-1} \leq |x|^{-n-1}$$

that for the second of the two integrals—which we denote by I_2—into which we can divide the right-hand side of (99), we have

$$|I_2| = \left| \int_\Gamma e^{x-\xi} \xi^{-1} G(\xi) \Phi(\xi)\, d\xi \right| \leq \frac{MN}{|x|^{n+1}} \int_\Gamma |e^{x-\xi}| \cdot |d\xi|$$

But the modulus of an exponential term is the exponential of its real part; therefore writing $\xi = \xi_1 + i\xi_2$ and denoting by θ the angle made by the tangent to the path Γ with the real axis,[*] we have

$$|d\xi| = \frac{d\xi_1}{|\cos \theta|} \leq \frac{d\xi_1}{\sin \eta}$$

whence follows

$$|I_2| \leq \frac{MN}{|x|^{n+1} \sin \eta} \int_\Gamma e^{\mathscr{R}x - \xi_1}\, d\xi_1 = \frac{MN}{|x|^{n+1} \sin \eta} \left[-e^{\mathscr{R}x - \xi_1} \right]_{\xi_1 = \mathscr{R}x}^{\xi_1 = +\infty}$$

i.e.

$$|I_2| \leq \frac{MN}{\sin \eta} |x|^{-n-1}$$

We now consider the first of the two integrals—we denote it by I_1—into which the right-hand side of (99) can be divided; here, by the inequalities (106), we have in case (2)—with obvious modifications in cases (1) and (3)

$$|I_1| = \left| \int_\Gamma \xi^{-2} F(\xi) \Phi(\xi)\, d\xi \right| \leq MN \int_x^{x'} |\xi|^{-n-2}\, d\xi + MN \int_{x'}^{\infty} |\xi|^{-n-2}\, |d\xi|$$

$$\leq \frac{MN}{|x|^{n+2}} \cdot \frac{\pi}{2} |x| + MN \left[\frac{-|\xi|^{-n-1}}{n+1} \right]_{\xi = x'}^{\xi = \infty}$$

whence

$$|I_1| \leq \frac{MN}{|x|^{n+1}} \left(\frac{\pi}{2} + \frac{1}{n+1} \right) \leq MN |x|^{-n-1} \left(1 + \frac{\pi}{2} \right)$$

---

[*] On the radial part of Γ this angle will coincide with the argument of x or with that of x' or that of x''.

We therefore conclude that, in all cases,

$$|T[\Phi]| \leq \frac{MN}{|x|^{n+1}}\left(1+\frac{\pi}{2}+\frac{1}{\sin\eta}\right)$$

which is exactly (105) provided

(107) $$K = M\left(1+\frac{\pi}{2}+\frac{1}{\sin\eta}\right)$$

which is a constant independent of the special function Φ considered.

We now use this general theorem on linear functional transformations to establish that the successive approximations (101) tend to a *function-limit* $Z(x)$ as $n \to \infty$—in fact, that this convergence is *uniform* within the domain D (see figure 26 translated into the x-plane) defined by the inequalities

(108) $$-\pi+\eta \leq \arg x \leq \pi-\eta, \qquad |x| \geq K^*$$

where K^* denotes any number larger that the greater of the two preceding numbers K and R^*.

$z_n(x)$ is in fact the sum to $(n+1)$ terms of the series

(109) $$z_0(x)+[z_1(x)-z_0(x)]+[z_2(x)-z_1(x)]+[z_3(x)-z_2(x)]+\ldots$$

Let x be any point in the domain D; with x we may associate a path of integration Γ such that for the corresponding transformation T the property '(104) implies (105)' is valid, where K has the value given by (107); applying this property to the successive terms in (109) and employing (103), we derive successively

$$|z_1(x)-z_0(x)| = |T[1]| \leq \frac{K}{|x|}$$

$$|z_2(x)-z_1(x)| = |T[z_1(x)-z_0(x)]| \leq \left(\frac{K}{|x|}\right)^2$$

$$|z_3(x)-z_2(x)| = |T[z_2(x)-z_1(x)]| \leq \left(\frac{K}{|x|}\right)^3$$

$$\cdots\cdots\cdots\cdots\cdots\cdots$$

and in general

(110) $$|z_n(x)-z_{n-1}(x)| \leq \left(\frac{K}{|x|}\right)^n \leq \left(\frac{K}{K^*}\right)^n \qquad (n=1,2,3,\ldots)$$

§ 49] An application of the method of successive approximations

But this shows that the series (109) is majorized by the geometric series $\Sigma(K/K^*)^n$ which is convergent since $K^* > K$; therefore the series (109) is absolutely and uniformly convergent within the domain D, i.e. within D the successive approximations $z_0, z_1, z_2, \ldots$ tend uniformly to a defined function-limit $Z(x)$, the sum of series (109).

50. Asymptotic integration of the reduced equation

Having established the uniform convergence of the successive approximations to a function-limit $Z(x)$ we can easily complete the 'asymptotic integration' of the reduced equation (95) by showing that

(I) $Z(x)$ satisfies the integral equation (97) and therefore also the differential equation (95) in the domain D.

(II) $Z(x)$ actually tends to the limit 1 as $x \to \infty$.

(III) $Z(x)$ may be represented asymptotically (in the sense defined earlier) by a series of the form

$$c_0 + \frac{c_1}{x} + \frac{c_2}{x^2} + \ldots$$

To verify property I we write

$$Z(x) - z_n(x) = R_n(x)$$

The recurrence formula (102) may now be written in the form

$$Z(x) - R_{n+1}(x) = 1 + T[Z(\xi) - R_n(\xi)] = 1 + T[Z(\xi)] - T[R_n(\xi)]$$

from which follows

$$Z(x) - 1 - T[Z(\xi)] = R_{n+1}(x) - T[R_n(\xi)]$$

But as the sequence of successive approximations is uniformly convergent, given any positive number ε as small as we please we can find an n_0 such that for $n > n_0$

$$|R_n(x)| < \varepsilon \qquad |R_{n+1}(x)| < \varepsilon$$

and therefore

$$|T[R_n(\xi)]| < \frac{K\varepsilon}{|x|} < \varepsilon$$

Hence for $n > n_0$ it follows that

$$|Z(x) - 1 - T[Z(\xi)]| < 2\varepsilon$$

which, in view of the arbitrariness of ε and the fact that the left-hand side is independent of n, implies

$$Z(x) - 1 - T[Z(\xi)] = 0$$

i.e. the function $Z(x)$ satisfies the integral equation (100).

Property II can be verified as briefly; for, by (110),

$$|Z(x)| \leq 1 + \sum_{n=1}^{\infty} |z_n(x) - z_{n-1}(x)| \leq \sum_{n=0}^{\infty} \left(\frac{K}{K^*}\right)^n = \frac{K^*}{K^* - K}$$

Thus, since $Z(x)$ satisfies the equation (100), it follows that

$$|Z(x) - 1| = |T[Z(\xi)]| \leq \frac{KK^*}{K^* - K} \cdot \frac{1}{|x|}$$

This is more than sufficient to imply that the limit of $Z(x)$ as $x \to \infty$ is 1; in fact we derive from the above inequality that

(111) $$Z(x) = 1 + O(x^{-1})$$

To verify property III we begin with the proof of a second general theorem on the functional transformation T:

If the function $\Phi(x)$ (on which the transformation operates) may be represented in the domain D by an asymptotic formula of the type

(112) $$\Phi(x) = \frac{k_m}{x^m} + \frac{k_{m+1}}{x^{m+1}} + \ldots + \frac{k_{n-1}}{x^{n-1}} + O(x^{-n}) \qquad (0 \leq m < n)$$

its transform $\phi(x) = T[\Phi(\xi)]$ can be represented within the same domain by a similar formula in which the exponents are all increased by 1, viz.

(113) $$\phi(x) = \frac{k^*_{m+1}}{x^{m+1}} + \frac{k^*_{m+2}}{x^{m+2}} + \ldots + \frac{k^*_n}{x^n} + O(x^{-(n+1)})$$

where $k^*_{m+1}, k^*_{m+2}, \ldots, k^*_n$ are certain constants depending on $k_m, k_{m+1}, \ldots, k_{n-1}$.

From the property of the transformation T established earlier that '(104) implies (105)' it follows that

$$T[O(x^{-n})] = O(x^{-(n+1)})$$

The proof therefore depends on showing that the result is valid for the transformation of the rational function

$$\Phi^*(x) = \frac{k_m}{x^m} + \frac{k_{m+1}}{x^{m+1}} + \ldots + \frac{k_{n-1}}{x^{n-1}}$$

On transformation of $\Phi^*(x)$ we write

$$\phi^*(x) = \phi_1(x) + \phi_2(x)$$

where

$$\phi_1(x) = \int_\Gamma \xi^{-2} F(\xi) \Phi^*(\xi) \, d\xi, \qquad \phi_2(x) = \int_\Gamma e^{x-\xi} \xi^{-1} G(\xi) \Phi^*(\xi) \, d\xi$$

i.e.

(114)
$$\begin{cases} \phi_1(x) = \int_\Gamma \left(\frac{\beta_1}{\xi^{m+2}} + \frac{\beta_2}{\xi^{m+3}} + \ldots \right) d\xi \\ \phi_2(x) = \int_\Gamma e^{x-\xi} \left(\frac{\gamma_1}{\xi^{m+1}} + \frac{\gamma_2}{\xi^{m+2}} + \ldots \right) d\xi \end{cases}$$

where $\beta_1, \beta_2, \ldots$ and $\gamma_1, \gamma_2, \ldots$ denote respectively the successive coefficients in the power-series expansions in $1/\xi$ of the two analytic functions $\xi^{-2} F(\xi) \Phi^*(\xi)$ and $\xi^{-1} G(\xi) \Phi^*(\xi)$ which have zeros of orders $m+2$ and $m+1$ respectively at the point $\xi = \infty$.

From the first of equations (114) it follows at once that there is an asymptotic formula for $\phi_1(x)$ of the type (113), in fact an ordinary expansion in series of the form

$$\phi_1(x) = \frac{\beta_1/(m+1)}{x^{m+1}} + \frac{\beta_2/(m+2)}{x^{m+2}} + \ldots$$

The function $\phi_2(x)$ may be written in the form

$$\phi_2(x) = \sum_{\nu=m+1}^{\infty} \gamma_{\nu-m} I_\nu(x)$$

where

$$I_\nu(x) = \int_\Gamma e^{x-\xi} \xi^{-\nu} d\xi$$

On integration by parts this gives

$$I_\nu(x) = [-e^{x-\xi} \xi^{-\nu}]_x^\infty - \nu \int_\Gamma e^{x-\xi} \xi^{-\nu-1} d\xi = x^{-\nu} - \nu I_{\nu+1}(x)$$

from which follow immediately

$$I_\nu(x) = x^{-\nu} - \nu x^{-(\nu+1)} + \nu(\nu+1) I_{\nu+2}(x)$$
$$I_\nu(x) = x^{-\nu} - \nu x^{-(\nu+1)} + \nu(\nu+1) x^{-(\nu+2)} - \nu(\nu+1)(\nu+2) I_{\nu+3}(x)$$
$$\cdot \quad \cdot \quad \cdot \quad \cdot \quad \cdot \quad \cdot \quad \cdot \quad \cdot \quad \cdot \quad \cdot \quad \cdot \quad \cdot$$
$$I_\nu(x) = x^{-\nu} - \nu x^{-(\nu+1)} + \nu(\nu+1) x^{-(\nu+2)} - \ldots$$
$$+ (-1)^{n-\nu} \nu(\nu+1) \ldots (n-1) [x^{-n} - n I_{n+1}(x)]$$

But these formulæ imply*

$$I_{n+1}(x) = O(x^{-(n+1)})$$

Thus for $\nu \leq n$ we may write

$$I_\nu(x) = x^{-\nu} - \nu x^{-(\nu+1)} + \nu(\nu+1) x^{-(\nu+2)} - \ldots$$
$$+ (-1)^{n-\nu} \nu(\nu+1) \ldots (n-1) x^{-n} + O(x^{-(n+1)})$$

and this is sufficient to assert that the function $\phi_2(x)$—and therefore also $\phi^*(x)$—may be represented by an asymptotic formula of the type (113), as we wished to show.

We now use this theorem to derive asymptotic forms for successive terms of the series (109), at least as far as the terms of order x^{-n-1} (with fixed n). We start with $z_1(x) - z_0(x) = T[1]$, for which case the equations (112)–(113) (here putting $m = 0$, $k_0 = 1$, $k_1 = k_2 = k_3 = \ldots = k_{n-1} = 0$) give

$$z_1(x) - z_0(x) = z_1(x) - 1 = \frac{k_1'}{x} + \frac{k_2'}{x^2} + \frac{k_3'}{x^3} + \ldots + \frac{k_n'}{x^n} + O(x^{-n-1})$$

where $k_1', k_2', \ldots, k_n'$ are suitable constants (depending on the coefficients of the functions F and G). We apply the preceding result a second time, now with $m = 1$, obtaining (writing down only the first n terms in this expansion)

$$z_2(x) - z_1(x) = T[z_1(\xi) - z_0(\xi)] = \frac{k_2''}{x^2} + \frac{k_3''}{x^3} + \ldots + \frac{k_n''}{x^n} + O(x^{-n-1})$$

where $k_2'', k_3'', \ldots, k_n''$ are new suitable constants. Similarly by further applications of the same result we find that

$$z_3(x) - z_2(x) = T[z_2(\xi) - z_1(\xi)] = \frac{k_3'''}{x^3} + \ldots + \frac{k_n'''}{x^n} + O(x^{-n-1})$$

---

* In fact the first relation is sufficient; for it is easily seen, by taking the modulus of the integral $I_{\nu+1}$ and observing that $|e^{x-\xi}| \leq 1$ on Γ, that $I_{\nu+1}$ is $O(x^{-\nu})$; now using this result and putting $\nu = n + 1$ into the first relation we derive the stated property.

§ 50]   Asymptotic integration of the reduced equation

and so on; ultimately

$$z_n(x) - z_{n-1}(x) = T[z_{n-1}(\xi) - z_{n-2}(\xi)] = \frac{k_n^{(n)}}{x^n} + O(x^{-n-1})$$

Now summing the preceding formulæ and putting

$$k_1' = c_1, \quad k_2' + k_2'' = c_2, \quad k_3' + k_3'' + k_3''' = c_3, \quad \ldots$$
$$k_n' + k_n'' + k_n''' + \ldots + k_n^{(n)} = c_n$$

we obtain the formula

(115) $$z_n(x) = 1 + \frac{c_1}{x} + \frac{c_2}{x^2} + \ldots + \frac{c_n}{x^n} + O(x^{-n-1})$$

in which the constants $c_1, c_2, \ldots, c_n$ are *stable* in the sense that in the transition from (115) to the corresponding formula for $z_{n+1}(x)$ they remain unaltered as

$$z_{n+1}(x) - z_n(x) = T\left[\frac{k_n^{(n)}}{\xi^n} + O(\xi^{-n-1})\right] = O(x^{-n-1})$$

Also, as the successive terms of the series (109) are $O(x^{-n-1})$, it follows immediately from (115) that

(116) $$Z(x) = 1 + \frac{c_1}{x} + \frac{c_2}{x^2} + \ldots + \frac{c_n}{x^n} + O(x^{-n-1})$$

where $c_1, c_2, \ldots$ are suitable constants, i.e.

(116*) $$Z(x) \sim 1 + \frac{c_1}{x} + \frac{c_2}{x^2} + \ldots$$

as n is arbitrary.

51. Conclusion and further comments

The results obtained in the preceding sections, when applied to the equation

(117) $$\frac{d^2 y}{dx^2} + \left(a_0 + \frac{a_1}{x} + \frac{a_2}{x^2} + \ldots\right)\frac{dy}{dx} + \left(b_0 + \frac{b_1}{x} + \frac{b_2}{x^2} + \ldots\right) y = 0$$

which was our starting point, allow us to reach some important conclusions to which reference has already been made in § 28. In view of the transformations given in (89) and (94) these results may be stated as follows:

If the characteristic equation

(118) $$\alpha^2 + a_0\alpha + b_0 = 0$$

has two distinct roots α_1 and α_2, and if

(119) $$\rho_1 = -\frac{a_1\alpha_1 + b_1}{2\alpha_1 + a_0}, \quad \rho_2 = -\frac{a_1\alpha_2 + b_1}{2\alpha_2 + a_0}$$

equation (117) possesses two linearly independent integrals* y_1 and y_2 which within the neighbourhoods D_1 and D_2 of the point at infinity to be defined presently, may be represented asymptotically in the sense of Poincaré by the series

(120) $$\begin{cases} y_1(x) \sim e^{\alpha_1 x} x^{\rho_1} \left(1 + \frac{c'_1}{x} + \frac{c'_2}{x^2} + \ldots\right) \\ y_2(x) \sim e^{\alpha_2 x} x^{\rho_2} \left(1 + \frac{c''_1}{x} + \frac{c''_2}{x^2} + \ldots\right) \end{cases}$$

To define the neighbourhoods D_1 and D_2 we first write for convenience

(121) $$\arg(2\alpha_1 + a_0) = \beta_1, \quad \arg(2\alpha_2 + a_0) = \beta_2$$

The neighbourhoods D_1 and D_2 are now identified with the domains defined respectively by the inequalities

(122) $$\begin{cases} -\beta_1 + \eta_1 \leq \arg x \leq 2\pi - \beta_1 - \eta_1, & |x| \geq K_1^* \\ -\beta_2 + \eta_2 \leq \arg x \leq 2\pi - \beta_2 - \eta_2, & |x| \geq K_2^* \end{cases}$$

where η_1 and η_2 denote two positive numbers as small as we please, and K_1^* and K_2^* two sufficiently large numbers (depending on η_1 and η_2 respectively).

In briefer but less precise terms—the domains D_1 and D_2 are the vicinities of the point at infinity from which have been removed the points nearest to the half-lines on which respectively $\arg x = -\arg(2\alpha_1 + a_0)$ and $\arg x = -\arg(2\alpha_2 + a_0)$.**

---

* That the two integrals are linearly independent is seen by considerations similar to those used on p. 226 in the Fuchs' case.

** These general considerations may be applied in particular to the reduced equation (95) (in which $a_0 = -1$, $a_1 = A_0$, $b_0 = b_1 = 0$, and consequently $\alpha_1 = 0$, $\alpha_2 = 1$; $\rho_1 = 0$, $\rho_2 = -A_0$; $\beta_1 = \pi$, $\beta_2 = 0$); these yield the asymptotic representation (116*) (valid except for values of x on the negative real axis) and in addition the similar representation

(116**) $$Z^*(x) \sim e^x x^{-A_0} \left(1 + \frac{c_1^*}{x} + \frac{c_2^*}{x^2} + \ldots\right)$$

(valid except for values of x on the positive real axis) which asymptotically represents a new particular integral of the equation, linearly independent of the preceding one.

What happens if the two roots of the characteristic equation coincide? It is easily seen that in this case logarithmic terms appear; while in fact the integral y_1 is still represented by an asymptotic series of the first type in (120), the other integral y_2 is now represented by a series of the form

$$(123) \qquad y_2(x) \sim A y_1(x) \log x + e^{\alpha_1 x} x^{\rho_1} \left(k_0 + \frac{k_1}{x} + \frac{k_2}{x^2} + \cdots \right)$$

To establish this result however we require results with which it is impossible to deal here.*

We consider briefly the determination of the constants $c'_1, c'_2, \ldots$, etc. which appear in formulæ (120). These can be found by carrying out in the reverse direction the previous calculations. For example, consider c_1 which represents c'_1 or c''_1 according as it appears in the formula for y_1 or in that for y_2; the constant c_1 is simply the coefficient k'_1 of $1/x$ in the expression for $T[1]$ and as such coincides (page 252) with the sum

$$\beta_1 + \gamma_1 = F(\infty) + G(\infty)$$

which, by using the equations (98), equals $-f(\infty) - g(\infty) + f(\infty) = -g(\infty)$, i.e.

$$(124) \qquad c_1 = -B_0$$

in the notation used in (96). The method however is long and tedious.

This laborious process may be avoided by using the usual method of undetermined coefficients, i.e. by substituting the expressions (120), or (123) if applicable, directly into the given equation or—as is usually preferable to avoid heavy calculation—by substituting the expression

$$(116^*) \qquad z(x) \sim 1 + \frac{c_1}{x} + \frac{c_2}{x^2} + \cdots$$

into the reduced equation (95). For this we require asymptotic expressions for $z'(x)$ and $z''(x)$, and we write

$$(116') \qquad z'(x) \sim -\frac{c_1}{x^2} - \frac{2c_2}{x^3} - \frac{3c_3}{x^4} - \cdots$$

$$(116'') \qquad z''(x) \sim \frac{1 \cdot 2 c_1}{x^3} + \frac{2 \cdot 3 c_2}{x^4} + \frac{3 \cdot 4 c_3}{x^5} + \cdots$$

---

* For study of this see W. STERNBERG: 'Über die asymptotische Integration von Differentialgleichungen', *Math. Ann.*, **81**, 119–86 (1926). This is a comprehensive work in which methods similar to those used here are applied to the study (for equations of any order) of the cases mentioned in the note on p. 241 and extended to the theory of partial differential equations. See also W. I. TRJITZINSKI: *Acta Math.*, **62**, 167–226 (1934); *Trans. Amer. Math. Soc.*, **37**, 80–146 (1935).

The question immediately arises whether this procedure is permissible—as asymptotic relations in general may be integrated but not differentiated.

It can be claimed that the calculations are *formal calculations* but that in itself does not guarantee the correctness of the results. In this case however the difficulty may be removed by proving that (116*) can be differentiated not only once but as often as we please, i.e. by showing that the derivatives of z may be represented asymptotically by formulæ of the type (116*). Once we have established that $z'(x)$ can be represented by a series of the type

$$(125) \qquad z'(x) \sim k_0 + \frac{k_1}{x} + \frac{k_2}{x^2} + \ldots$$

it will follow immediately that the coefficients in this representation must be those appearing in (116'), i.e.

$$k_0 = k_1 = 0, \qquad k_2 = -c_1, \qquad k_3 = -2c_2, \ldots$$

for, if not, then on integrating (125) (which is permissible, provided attention be paid to the first two terms) we would have a second asymptotic representation of $z(x)$ different from (116*), contrary to the established uniqueness of such representations. Further, it is sufficient to show that the first derivative of z may be represented asymptotically as in (125)—for once we have established this result the representability of z'' follows at once from the differential equation (95) which may be written as

$$z'' = [1 - x^{-1} f(x)] z' - x^{-2} g(x) z$$

and similarly for successive derivatives, by differentiating both sides of this equality.

To show that z' can be represented by an asymptotic series of the type (125) we differentiate the integral equation (97), giving

$$z'(x) = \int_x^\infty e^{x-\xi} \xi^{-1} G(\xi) z(\xi) \, d\xi - [x^{-2} F(x) + x^{-1} G(x)] z(x)$$

Addition and subtraction of $\xi^{-2} F(\xi) z(\xi)$ within the integrand now gives

$$(126) \qquad z'(x) = -[x^{-2} F(x) + x^{-1} G(x)] z(x) + T[z(\xi)] - \int_x^\infty \xi^{-2} F(\xi) z(\xi) \, d\xi$$

which, in view of the results already obtained and of (116*), yields a representation of the type (125) for the function $z'(x)$ with, in fact, $k_0 = k_1 = 0$.

The formulæ (116') and (116'') are therefore justified; now substituting these expressions together with (116*) into the right-hand side of equation (95), collecting together all the terms in the same power of $1/x$ and equating to zero the coefficients of successive powers, we derive the following linear equations

$$
(127) \quad \begin{cases} 1 \cdot c_1 + B_0 = 0 \\ 2 \cdot c_2 + (1 \cdot 2 - A_0 + B_0) c_1 + B_1 = 0 \\ 3 \cdot c_3 + (2 \cdot 3 - 2A_0 + B_0) c_2 + (B_1 - A_1) c_1 + B_2 = 0 \\ \cdot \cdot \cdot \cdot \cdot \cdot \cdot \cdot \cdot \cdot \cdot \cdot \cdot \cdot \cdot \cdot \cdot \\ (n+1) c_{n+1} - \sum_{m=1}^{n} (mA_{n-m} - B_{n-m}) c_m + n(n+1) c_n + B_n = 0 \\ \cdot \cdot \cdot \cdot \cdot \cdot \cdot \cdot \cdot \cdot \cdot \cdot \cdot \cdot \cdot \cdot \cdot \end{cases}
$$

The first of these immediately yields the value (124) of c_1, and successive equations determine in turn the other c's. [It should be noted that the zero is represented asymptotically by the series

$$0 + \frac{0}{x} + \frac{0}{x^2} + \ldots]$$

In conclusion we wish to emphasize that one of the most important results proved by Sternberg in the work cited earlier is that *in all cases*—even for linear equations of higher order, or for equations whose coefficients have poles, or whose characteristic equations have multiple roots, etc.—the results obtained by *formal* calculations similar to those above can always be justified in terms of asymptotic representations, i.e. they lead to correct asymptotic representations of the unknown functions and of their derivatives.

52. Applications to confluent hypergeometric functions and to Bessel functions

The preceding results concerning the asymptotic behaviour of differential equations find elegant applications in the theory of *confluent hypergeometric functions*, which include as particular cases the Bessel functions and many other important functions.

Such functions may be considered as limiting cases of the general hypergeometric function, the Riemann P-function, when two of the three singular points tend to *coalesce* at the point at infinity. Using the transformations which in § 47 allowed us to reduce the number of parameters in the P-function to only three, we consider the function

(128) $$P\left\{\begin{matrix} 0 & \alpha_2 & \alpha_3 & \\ 0 & 0 & a & x \\ 1-c & c-a-b & b & \end{matrix}\right\}$$

which satisfies the linear second-order equation obtained immediately from the Papperitz equation (69) by putting $a = 0$, $b = \alpha_2$, $c = \alpha_3$, and $\alpha = 0$, $\alpha' = 1-c$, $\beta = 0$, $\beta' = c-a-b$, $\gamma = a$, $\gamma' = b$; this equation is

$$y'' + \left[\frac{c}{x} - \frac{(\alpha_3-\alpha_2)b}{\alpha_2\alpha_3}\frac{\alpha_2}{\alpha_2-x}\frac{\alpha_3}{\alpha_3-x} - \frac{a-c+1}{\alpha_2-x} - \frac{1-a}{\alpha_3-x}\right]y'$$
$$- a\frac{(\alpha_3-\alpha_2)b}{\alpha_2\alpha_3}\frac{\alpha_2}{\alpha_2-x}\left(\frac{\alpha_3}{\alpha_3-x}\right)^2\frac{1}{x}y = 0$$

Now, while a and c remain fixed, let α_2, α_3 and b tend to infinity in such a way that

$$\lim\frac{(\alpha_3-\alpha_2)b}{\alpha_2\cdot\alpha_3} = 1$$

as, for example, occurs if we write $\alpha_2 = b/2$, $\alpha_3 = b$ and allow b to tend to infinity in any way. Under these hypotheses we find

$$\lim\frac{\alpha_2}{\alpha_2-x} = \lim\frac{\alpha_3}{\alpha_3-x} = 1, \quad \lim\frac{a-c+1}{\alpha_2-x} = \lim\frac{1-a}{\alpha_3-x} = 0$$

and the preceding differential equation tends to the 'equation-limit', viz. the *confluent* equation

(129) $$xy'' + (c-x)y' - ay = 0$$

This last section will be devoted to the study of equation (129), and particularly of its solution by means of asymptotic series.

The equation (129), which is substantially identical with the Laguerre equation discussed in § 37 and to which it reduces on writing $c = \alpha+1$, $a = -\lambda$, is one of the most important in analysis as its solutions include most of the special functions which arise in pure and applied mathematics. For example, among these, in addition to the Laguerre polynomials and special cases of these, are Bessel functions, as can be seen easily since the Bessel equation

$$\frac{d^2z}{dx^2} + \frac{1}{x}\frac{dz}{dx} + \left(1 - \frac{v^2}{x^2}\right)z = 0$$

§ 52] Applications to confluent hypergeometric functions and Bessel functions

is transformed into

$$\xi \frac{d^2 y}{d\xi^2} + (1 + 2\nu - \xi) \frac{dy}{d\xi} - (\tfrac{1}{2} + \nu) y = 0$$

on putting

$$y = x^{-\nu} e^{ix} z, \qquad \xi = 2ix$$

This new equation is the particular case of (129) for which

$$a = \nu + \tfrac{1}{2}, \qquad c = 2a = 2\nu + 1$$

In other words, if we denote by $Z_\nu(x)$ a solution of the Bessel equation and by $\mathscr{F}(a, c; x)$ a solution of (129) (i.e. a *confluent hypergeometric function*) we obtain the result

(130) $$Z_\nu(x) = x^\nu e^{-ix} \mathscr{F}(\nu + \tfrac{1}{2}, 2\nu + 1; 2ix)$$

The only singular points of the differential equation (129) are the origin and the point at infinity; the Fuchs' conditions are satisfied evidently at the origin but not at the point at infinity, as is seen on dividing the equation by the leading coefficient—for then it is obvious that the coefficients of y' and of y do not possess the required zeros of first and second orders respectively for $x = \infty$ (cf. p. 229).

The most important confluent hypergeometric function is derived by substituting into (129) the power series $\Sigma k_n x^n$ for y; it is easily found that this series can satisfy the equation only if the recurrence relations

$$k_n = \frac{a + n - 1}{n(c + n - 1)} k_{n-1}$$

hold (we exclude here the cases in which c is a negative integer or zero); this leads to the *Kummer series*★

(131) $$\Phi(a, c; x) = 1 + \frac{a}{c} \frac{x}{1!} + \frac{a(a+1)}{c(c+1)} \frac{x^2}{2!} + \cdots$$

$$+ \frac{a(a+1) \cdots (a+n-1)}{c(c+1) \cdots (c+n-1)} \frac{x^n}{n!} + \cdots$$

with *infinite radius of convergence*. The function $\Phi(a, c; x)$ is in fact the limiting case of the hypergeometric function $F(a, b; c; x/b)$ when $b \to \infty$ (see page 237).

★ The symbol $_1F_1(a; c; x)$ is frequently used.

Some particular cases of $\Phi(a, c; x)$ are

(132)
$$\begin{cases} \Phi(a, a; x) = e^x, \quad \Phi(\tfrac{1}{2}, \tfrac{3}{2}; -x^2) = \dfrac{1}{x}\int_0^x e^{-t^2}\,dt \\ \Phi(1, 1+\alpha; x) = \alpha x^{-\alpha} e^x \int_0^x e^{-t} t^{\alpha-1}\,dt \\ \Phi(-n, \alpha+1; x) = \binom{\alpha+n}{n} L_n^{(\alpha)}(x), \quad \text{etc.} \end{cases}$$

Since equation (129) is transformed into another equation of the same type under the two transformations

(133) $$y = x^{1-c} y_1, \quad y = e^x y_1$$

we obtain the two important relations

(134) $$\mathscr{F}(a, c; x) = x^{1-c}\mathscr{F}(a-c+1, 2-c; x)$$

and

(135) $$\mathscr{F}(a, c; x) = e^x \mathscr{F}(c-a, c; -x)$$

which when applied successively yield

(136) $$\mathscr{F}(a, c; x) = e^x x^{1-c}\mathscr{F}(1-a, 2-c; -x)$$

It follows from (134) that *provided that c is not integral* the general integral of (129) may be written as

(137) $$y = C_1 \Phi(a, c; x) + C_2 x^{1-c}\Phi(a-c+1, 2-c; x)$$

where C_1 and C_2 are arbitrary constants, and Φ is the *complete* transcendental function defined by (131).

Also, formulæ (135) applied to the function Φ produces the important *Kummer formula*

(138) $$\Phi(a, c; x) = e^x \Phi(c-a, c; -x)$$

After this brief summary of general results on the confluent functions*

---

* For greater detail see F. TRICOMI: *Funzioni ipergeometriche confluenti* (Rome, Cremonese, 1954) and " Fonctions hypergéometriques confluentes," *Mém. Sciences Math.* No. 140, Paris, 1959; also E. T. WHITTAKER-G. N. WATSON (85), Chapter XVI, in which the functions studied, in place of the functions Φ, are the functions

$$M_{k,m}(x) = x^{1/2+m} e^{-x/2}\Phi(1/2 + m - k, 2m + 1; x)$$

and the functions $W_{k,m}(x)$ which (apart from exceptional cases) are linear combinations of $M_{k,m}(x)$ and $M_{k,-m}(x)$. These functions are also studied in the books of H. BUCHHOLZ: *Die konfluente hypergeometrische Funktion* (Berlin, Springer, 1953) and L. J. SLATER: *Confluent Hypergeometric Functions* (Cambridge University Press, 1960).

§ 52] Applications to confluent hypergeometric functions and Bessel functions

we try to determine the asymptotic behaviour of such functions for large values of $|x|$ by applying the methods of the last sections.

We start by reducing the equation (129) to the form (95); using a transformation similar to the first of (133) we write

$$y = x^{-a} z$$

Equation (129) then becomes

(139) $$z'' - \left[1 - \frac{1}{x}(c-2a)\right]z' + \frac{1}{x^2}a(a-c+1)z = 0$$

which is of the form (95) with

$$f(x) \equiv c-2a, \qquad g(x) \equiv a(a-c+1)$$

whence

$$A_0 = c-2a, \quad B_0 = a(a-c+1), \quad A_1 = B_1 = A_2 = B_2 = \ldots = 0$$

We may therefore assert that (139) possesses an integral which can be represented asymptotically by the series $1 + c_1/x + c_2/x^2 + \ldots$, i.e. that (139) possesses an integral $z_1{}^*$ such that

$$z_1(x) \sim 1 + \frac{c_1}{x} + \frac{c_2}{x^2} + \ldots$$

(for all values of x except negative real values) where the constants $c_1, c_2, \ldots$ are calculated by the linear equations (127); in this case these equations are

$$c_1 + a(a-c+1) = 0$$
$$(n+1)c_{n+1} + [n(n+1) - (c-2a)n + a(a-c+1)]c_n = 0 \qquad (n = 1, 2, \ldots)$$

hence

$$c_{n+1} = -\frac{(n+a)(n+a-c+1)}{n+1} c_n$$

and consequently

$$c_n = (-1)^n \frac{a(a+1)\ldots(a+n-1)\cdot(a-c+1)(a-c+2)\ldots(a-c+n)}{n!}$$

$$= (-1)^n \binom{-a}{n}\binom{c-a-1}{n} n!$$

---

* In the author's books cited on the previous page an integral of this form is denoted by the symbol $\Psi(a, c; x)$. This integral is connected to the Whittaker function $W_{k,m}$ in a similar way to that in which $\Phi(a, c; x)$ is related to $M_{k,m}$.

We therefore find that in a neighbourhood of the point at infinity from which are excluded the points near the negative real axis

$$(140) \qquad y_1(x) \sim x^{-a} \sum_{n=0}^{\infty} (-1)^n \binom{-a}{n}\binom{c-a-1}{n}\frac{n!}{x^n}$$

To find an asymptotic series for another particular integral $y_2(x)$ linearly independent of the first integral we use (135), applying to the function $\mathscr{F}$ on the right-hand side of (135) the expansion just obtained for $y_1(x)$; this gives

$$(141) \qquad y_2(x) \sim e^x x^{a-c} \sum_{n=0}^{\infty} \binom{a-c}{n}\binom{a-1}{n}\frac{n!}{x^n}$$

where in this case the points near the positive real axis are excluded.

We therefore conclude that *any confluent hypergeometric function* $\mathscr{F}(a, c; x)$ *can be expanded asymptotically (except for real values of x) as*

$$\mathscr{F} \sim C_1 x^{-a} \sum_{n=0}^{\infty} (-1)^n \binom{-a}{n}\binom{c-a-1}{n}\frac{n!}{x^n} + C_2 e^x x^{a-c} \sum_{n=0}^{\infty} \binom{a-c}{n}\binom{a-1}{n}\frac{n!}{x^n}$$

where C_1 and C_2 are two constants depending on the special function considered.

In the important case of the Bessel functions we therefore find that *except for points on the imaginary axis*

$$Z_\nu(x) \sim x^\nu e^{-ix}\Bigg[C_1 (2ix)^{-\nu-\frac{1}{2}} \sum_{n=0}^{\infty} (-1)^n \binom{-\frac{1}{2}-\nu}{n}\binom{-\frac{1}{2}+\nu}{n}\frac{n!}{(2x)^n}(-i)^n$$
$$+ C_2 e^{2ix} (2ix)^{-\nu-\frac{1}{2}} \sum_{n=0}^{\infty} \binom{-\frac{1}{2}-\nu}{n}\binom{-\frac{1}{2}+\nu}{n}\frac{n!}{(2x)^n}(-i)^n \Bigg]$$

Now writing

$$i^n = e^{\frac{n\pi}{2}i}, \qquad (-i)^n = e^{-\frac{n\pi}{2}i}$$

and denoting by C_ν and γ_ν two new constants defined by

$$(2i)^{-\nu-\frac{1}{2}} C_1 = -\frac{C_\nu}{2i} e^{-\gamma_\nu i}, \qquad (2i)^{-\nu-\frac{1}{2}} C_2 = \frac{C_\nu}{2i} e^{\gamma_\nu i}$$

we derive by easy transformations that

$$Z_\nu(x) \sim \frac{C_\nu}{\sqrt{x}} \sum_{n=0}^{\infty} \binom{-\frac{1}{2}-\nu}{n}\binom{-\frac{1}{2}+\nu}{n}\frac{n!}{(2x)^n} \sin\left(x + \gamma_\nu - n\frac{\pi}{2}\right)$$

§ 52] *Applications to confluent hypergeometric functions and Bessel functions*   263

Separating the terms for which n is even ($n = 2m$) from those for which n is odd ($n = 2m+1$) and noting that

$$\sin\left(x+\gamma_\nu-2m\frac{\pi}{2}\right) = (-1)^m \sin(x+\gamma_\nu)$$

$$\sin\left[x+\gamma_\nu-(2m+1)\frac{\pi}{2}\right] = -(-1)^m \cos(x+\gamma_\nu)$$

and

$$n!\binom{-\tfrac{1}{2}-\nu}{n}\binom{-\tfrac{1}{2}+\nu}{n} = \frac{1}{n!}(-\tfrac{1}{2}-\nu)(-\tfrac{1}{2}+\nu)(-\tfrac{3}{2}-\nu)(-\tfrac{3}{2}+\nu)$$

$$\cdots \left(-\frac{2n-1}{2}-\nu\right)\left(-\frac{2n-1}{2}+\nu\right)$$

$$= \frac{1}{n!}[(\tfrac{1}{2})^2-\nu^2][(\tfrac{3}{2})^2-\nu^2] \cdots \left[\left(\frac{2n-1}{2}\right)^2-\nu^2\right]$$

$$= \frac{(-1)^n}{4^n n!}(4\nu^2-1^2)(4\nu^2-3^2) \cdots [4\nu^2-(2n-1)^2]$$

we obtain classic asymptotic expansion

(142) $$Z_\nu(x) \sim \frac{C_\nu}{\sqrt{x}} \sin(x+\gamma_\nu)\left[1 - \frac{A_2^{(\nu)}}{2!(8x)^2} + \frac{A_4^{(\nu)}}{4!(8x)^4} - \cdots\right]$$

$$+ \frac{C_\nu}{\sqrt{x}} \cos(x+\gamma_\nu)\left[\frac{A_1^{(\nu)}}{1!(8x)} - \frac{A_3^{(\nu)}}{3!(8x)^3} + \frac{A_5^{(\nu)}}{5!(8x)^5} - \cdots\right]$$

where*

(143) $$A_n^{(\nu)} = (4\nu^2-1^2)(4\nu^2-3^2)(4\nu^2-5^2) \cdots [4\nu^2-(2n-1)^2]$$

In particular, for the *Bessel function of the first kind* $J_\nu(x)$—from (142) follows the result

$$Z_\nu(x) = \frac{C_\nu}{\sqrt{x}} \sin(x+\gamma_\nu) + O(x^{-3/2})$$

which is simply the formula (49) of Chapter IV (page 153); therefore

$$C_\nu = \sqrt{\left(\frac{2}{\pi}\right)}, \quad \gamma_\nu = -\frac{\nu\pi}{2} + \frac{\pi}{4}$$

---

* For a generating function for the $A_n^{(\nu)}$ the reader is referred to *Higher Transcendental Functions*, v. III. This work has been previously cited on p. 240.

and consequently

(144) $$J_\nu(x) \sim \sqrt{\left(\frac{2}{\pi x}\right)} \cos\left(x - \frac{\nu\pi}{2} - \frac{\pi}{4}\right)\left[1 - \frac{A_2^{(\nu)}}{2!(8x)^2} + \cdots\right]$$
$$- \sqrt{\left(\frac{2}{\pi x}\right)} \sin\left(x - \frac{\nu\pi}{2} - \frac{\pi}{4}\right)\left[\frac{A_1^{(\nu)}}{1!(8x)} - \frac{A_3^{(\nu)}}{3!(8x)^3} + \cdots\right]$$

Figure 27 represents graphically the error functions $E_0(x)$, $E_1(x)$, and $E_2(x)$ arising in the evaluation of $J_0(x)$ by the series on the right-hand side of (144) when abbreviated to the first term, or to the term in x^{-1}, or to the

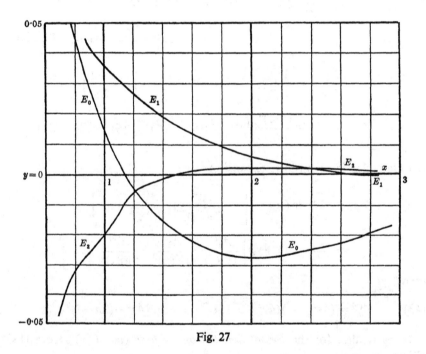

Fig. 27

term in x^{-2}, respectively; the values of x used lie between $x = 0.8$ and $x = 3$, and the ordinate scale is twenty times larger than the abscissa scale. This illustrates clearly how even two or three terms of the series give good approximations not only for large values of x but even for values greater than 2.

Bibliography

This index lists the books which have been cited in the text and others in which will be found useful supplementary material.

A. On Differential Equations

1. R. BELLMAN: *Stability Theory of Differential Equations* (New York, McGraw-Hill, 1954).
2. A. A. BENNETT, W. MILNE, H. BATEMAN: *Numerical Integration of Differential Equations* (New York, Dover, 1955).
3. L. BIANCHI: *Lezioni sulla teoria delle equazioni differenziali lineari. (Teoria di Fuchs-Riemann)* (Circolo Mat. Catania, 1924).
4. L. BIEBERBACH: *Theorie der Differentialgleichungen* (3rd ed., Berlin, Springer, 1930).
5. L. BIEBERBACH: *Theorie der gewöhnlichen Differentialgleichungen auf funktiontheoretischer Grundlage* (Berlin, Springer, 1953).
6. L. BIEBERBACH: *Einführung in die Theorie der Differentialgleichungen im reellen Gebiet* (Berlin, Springer, 1956).
7. M. BÔCHER: *Leçons sur les méthodes de Sturm* (Paris, Gauthier-Villars, 1917).
8. G. BOULIGAND (and J. DEVISME): *Lignes de niveau, lignes intégrales. Introduction à leur étude graphique* (Paris, Vuibert, 1937).
9. P. BOUTROUX: *Leçons sur les fonctions définies par les équations différentielles du premier ordre* (Paris, Gauthier-Villars, 1908).
10. E. A. CODDINGTON, N. LEVINSON: *Theory of Ordinary Differential Equations* (New York, McGraw-Hill, 1955).
11. L. COLLATZ: *Eigenwertprobleme und ihre numerische Behandlung* (2nd ed., Berlin, Springer, 1953).
12. L. COLLATZ: *Numerische Behandlung von Differentialgleichungen* (2nd ed., Berlin, Springer, 1955).
13. L. COLLATZ: *Differentialgleichungen für Ingenieure* (Wolfenbütteler Verlagsanstalt, 1949).
14. H. DULAC: *Curvas definidas por una ecuación diferencial de primer orden y de primer grado* (Madrid, C. Bermajo, 1933).
15. *Encyklopädie der mathematischen Wissenschaften.* Articles: II A, 4A (P. PAINLEVÉ) — II A, 4B (E. VESSIOT) — II A, 7A (M. BÔCHER) — II C, 2 (C. RUNGE, A. WILLERS) — II C, 11 (E. HILB, O. SZÁSZ) — III D, 8 (H. LIEBMANN).
16. A. R. FORSYTH: *Theory of Differential Equations* (6 vols., Cambridge U.P., 1890–1906); reprinted at New York (Dover, 1959).

17. P. FRANK, R. VON MISES: *Die Differential- und Integralgleichungen der Mechanik und Physik* (2nd ed., 2 vols., Braunschweig, Vieweg & S., 1930–1935).

18. K. O. FRIEDRICHS: *Advanced Ordinary Differential Equations* (New York University, 1956).

19. M. GOLOMB, M. SHANKS, *Elements of Ordinary Differential Equations* (New York, McGraw-Hill, 1950).

20. G. HOHEISEL: *Gewöhnliche Differentialgleichungen* (3rd ed., Berlin, W. de Gruyter [Samm. Göschen], 1938).

21. J. HORN: *Gewöhnliche Differentialgleichungen* (2nd ed., Berlin, W. de Gruyter, 1927).

22. W. HORT, A. THOMA: *Die Differentialgleichungen der Technik und Physik* (6th ed., Leipzig, Barth, 1954).

23. W. HUREWICZ: *Lectures on Ordinary Differential Equations* (New York, Wiley (and M.I.T.), 1958).

24. E. L. INCE: *Ordinary Differential Equations* (London, Longmans Green, 1927).

25. E. L. INCE: *Integration of Ordinary Differential Equations* (Edinburgh, Oliver and Boyd, 1952).

26. INTERN. CENTRO MATEM. ESTIVO (VARENNA): *Equazioni differenziali nonlineari*, (mimeographed, Rome, 1954).

27. E. KAMKE: *Differentialgleichungen reeller Funktionen* (3rd ed., Leipzig, 1956).

28. E. KAMKE: *Differentialgleichungen. Lösungsmethoden und Lösungen* (Leipzig, Akad. Verlagsgesell., 1942).

29. A. KNESCHKE: *Differentialgleichungen und Randwertprobleme* (Leipzig, Akad. Verlagsgesell., 1942).

30. A. I. LAPPO-DANILEVSKY: *Théorie des systèmes des équations différentielles linéaires* (New York, Chelsea, 1953).

31. S. LEFSCHETZ: *Lectures on Differential Equations* (Princeton U.P., 1948).

32. S. LEFSCHETZ (and others): *Contributions to the Theory of Non-linear Oscillations* (3 vols., Princeton U.P., 1950–57).

33. S. LEFSCHETZ: *Differential Equations: Geometric Theory* (New York, Interscience, 1957).

34. H. LEVY, E. A. BAGGOTT: *Numerical Solutions of Differential Equations* (New York, Dover, 1950).

35. H. LIEBMANN: *Lehrbuch der Differentialgleichungen* (Leipzig, v. Veit, 1901).

36. N. W. McLACHLAN: *Ordinary non-Linear Differential Equations in Engineering and Physical Sciences* (Oxford U.P., 1950).

37. *Mémorial des Sciences Mathématiques.* Fascicules: 28 (E. COTTON) — 48 (M. PETROVITCH) — 61 (H. DULAC) — 90 (W. I. TRJITZINSKI) — 97 (R. LAGRANGE) — 133 (G. HEILBRONN).

38. F. J. MURRAY, K. S. MILLER: *Existence Theorems for Ordinary Differential Equations* (New York Univ. Press, 1954).

39. A. D. MYSCHKIS: *Lineare Differentialgleichungen mit nacheilendem Argument* (Berlin, D. Verlag der Wiss., 1955).

40. V. V. NEMICKII, V. V. STEPANOV: *Qualitative Theory of Differential Equations* (Russian, 2nd ed., Moscow, Gos. Izd. Teckh. Teor. Lib., 1949; Translation, Princeton, 1958).

41. M. PETROVITCH: *Intégrales premières à restrictions* (Paris, Gauthier-Villars, 1929).
42. I. G. PETROVSKI: *Vorlesungen über die Theorie der gewöhnlichen Differentialgleichungen* (Leipzig, Teubner, 1954).
43. E. G. C. POOLE: *Introduction to the Theory of Linear Differential Equations* (Oxford U.P., 1936).
44. Proceedings of the Conference on Differential Equations at the University of Maryland (editors, T. B. DIAZ and L. E. PAYNE), 1956.
45. H. W. REDDICK: *Differential Equations* (2nd ed., New York, Wiley, 1950).
46. J. F. RITT: *Integration in Finite Terms (Liouville's Theory of Elementary Methods)* (New York, Columbia U.P., 1948).
47. G. SANSONE: *Equazioni differenziali nel campo reale* (3rd ed., 2 vols., Bologna, Zanichelli, 1948–49).
48. G. SANSONE, R. CONTI: *Equazioni differenziali nonlineari* (Rome, Cremonese, 1955).
49. F. SAUTER: *Differentialgleichungen der Physik* (Berlin, W. de Gruyter [Samm. Göschen], 1950).
50. L. SCHLESINGER: *Handbuch der Theorie der linearen Differentialgleichungen* (Leipzig, Teubner, 1895).
51. V. V. STEPANOV: *Lehrbuch der Differentialgleichungen* (Berlin, D. Verlag der Wiss., 1956).
52. *Studies in Differential Equations* (H. T. DAVIS, W. SCOTT, G. SPRINGER, D. RESCH) (Evanston, Northwest. Univ. Press, 1956).
53. E. C. TITCHMARSH, *Eigenfunction Expansions associated with Second Order Differential Equations* (Oxford U.P., 1946).
54. V. I. ZUBOV: *The Methods of A. M. Lyapounov and their Applications* (Russian, Moscow, Izdat. Leningrad Univ., 1957).

B. *On Related Topics*

55. A. A. ANDRONOV, C. E. CHAIKIN: *Theory of Oscillations* (Princeton U.P., 1951).
56. N. N. BOGOLIUBOV, I. A. MITROPOLSKI: *Asymptotic Methods in the Theory of Non-linear Oscillations* (Russian, Moscow, Gos. Izd. Teckh. Teor. Lit., 1955).
57. N. G. CETAEV: *Stability of Motion* (Russian, Moscow, Gos. Izd. Teckh. Teor. Lit., 1956).
58. R. V. CHURCHILL: *Modern Operational Mathematics in Engineering* (New York, McGraw-Hill, 2nd ed., 1958).
59. R. COURANT, D. HILBERT: *Methods of Mathematical Physics*, Vol. I (New York, 1953).
60. G. DOETSCH: *Theorie und Anwendung der Laplace-Transformation* (Berlin, Springer, 1937).
61. G. DOETSCH: *Handbuch der Laplace-Transformation*, Vols. I–III (Basel, Birkhäuser, 1950–56).
62. A. ERDÉLYI: *Asymptotic Expansions* (New York, Dover, 1956).

63. A. Ghizzetti: *Calcolo simbolico* (Bologna, Zanichelli, 1943).
64. E. Goursat: *Cours d'analyse mathématique* (4th ed., 3 vols., Paris, Gauthier-Villars, 1924–28).
65. C. Hayashi: *Forced Oscillations in Non-linear Systems* (Osaka, Nippon Print. Publ. Co., 1953).
66. E. Jahnke, F. Emde: *Tafeln höherer Funktionen* (6th ed., Stuttgart, Teubner, 1959); *Tables of Functions* (New York, Dover, 1945).
67. K. Knopp: *Theory and Application of Infinite Series* (2nd ed., Glasgow, Blackie, 1951).
68. N. Krylov, N. N. Bogoliubov: *Introduction to Non-linear Mechanics* (Free translation by S. Lefschetz, Princeton U.P., 1943).
69. W. Magnus, F. Oberhettinger: *Formeln und Sätze für die speziellen Funktionen der mathematischen Physik* (2nd ed., Berlin, Springer, 1948; in English, New York, Dover).
70. I. G. Malkin: *Methods of Liapounov and Poincaré in the Theory of Non-linear Oscillations* (Russian, Moscow-Leningrad, Gos. Izd., 1949).
71. I. G. Malkin: *Some Problems in the Theory of Non-linear Oscillations* (Russian, 1956); *Theory of Stability of Motion* (Russian, 1952) (both of these published by Moscow, Gos. Izd. Teckh. Teor. Lit.).
72. N. Minorsky: *Introduction to Non-linear Mechanics* (Ann Arbor, Mich., J. W. Edwards, 1947).
73. I. A. Mitropolski: *Non-stationary Processes in Non-linear Oscillating Systems* (Russian, Kiev, Izd. Akad. Nauk. Ukr. S.S.R., 1955).
74. E. Picard: *Traité d'analyse* (3rd ed., 3 vols., Paris, Gauthier-Villars, 1922–28).
75. M. Picone: *Appunti di analisi superiore* (2nd ed., Naples, Rondinella, 1946).
76. J. A. Serret: *Cours de calcul différentiel et intégral* (3rd ed., 2 vols., Paris, Gauthier-Villars, 1886; 4th ed. in German revised by G. Scheffers, Leipzig, Teubner, 1924).
77. J. J. Stoker: *Non-linear Vibrations in Mechanical and Electrical Systems* (New York, Interscience, 1950).
78. G. Szegö: *Orthogonal Polynomials*, Amer. Math. Soc. Coll. Publ. No. 23 (1939, revised 1948).
79. F. Tricomi: *Lezioni di analisi matematica* (7th ed., 2 vols., Padua, Cedam, 1956).
80. F. Tricomi: *Funzioni analitiche* (2nd ed., Bologna, Zanichelli, 1946; revised 1952).
81. F. Tricomi: *Serie ortogonali di funzioni* (Turin, Gheroni, 1948); *Vorlesungen über Orthogonalreihen* (Berlin, Springer, 1955).
82. F. Tricomi: *Integral Equations* (New York, Interscience, 1957).
83. C. de la Vallée Poussin: *Cours d'analyse infinitésimale* (8th ed., Paris, Gauthier-Villars, 1938).
84. (G. Vitali), G. Sansone: *Moderna teoria delle funzioni di variabile reale*, Vol. II: *Sviluppi in serie di funzioni ortogonali* (3rd ed., Bologna, Zanichelli, 1952).
85. E. T. Whittaker, G. N. Watson: *Modern Analysis* (4th ed., Cambridge U.P., 1927).

Author Index

Amerio, L., 85, 87
Andreev, A. F., 72
Andronov, A. A.-Chaikin, C.E., 87, 267
Appell, P.-Kampé de Ferier, J., 240
Ascoli, G, 149, 152, 158, 161, 162

Bailey, W. N., 240
Bateman Project, 240
Bautin, N., 66
Bellman, R., 161, 265
Bendixson, J., 39, 48, 52, 53, 72, 75, 84
Bennett, A. A.-Milne, W.-Bateman, H., 265
Bianchi, L., 265
Bieberbach, L., 87, 108, 209, 265
Bocher, M., 265
Bogoliubov, N. N.-Mitropolski, I. A., 267
Bouligand, G., 73, 265
Boutroux, P., 265
Buchholz, H., 260

Cetaev, N. G., 267
Churchill, R. V., 267
Coddington, E. A.-Levinson, N., 265
Collatz, L., 122, 265
Cotton, E., 266
Courant, R.-Hilbert, D., 136, 267

Davis, H. T.-Scott, W.-Springer, G.-Resch, D., 267
Devisme, J. (and Bouligand, G.), 265
Diaz, T. B.-Payne, L. E., 267
Dini, U., 141
Doetsch, G., 267
Dulac, H., 265, 266

Erdélyi, A., 177, 188, 267
Erdélyi, A.-Magnus, W.-Oberhettinger, F.-Tricomi, F., 240

Forster, H., 72
Forsyth, A. R., 2, 236, 265
Frank, P.-von Mises, R., 266
Friedrichs, K. O., 266

Frommer, M., 66, 72
Fubini, G., 142

Ghizzetti, A., 268
Golomb, M.-Shanks, M., 266
Goursat, E., 268
Graffi, D., 82
Greco, D., 138

Hamel, G., 136
Hartman, P., 92, 159
Hayashi, C., 268
Heilbron, G., 266
Hilb, E., 265
Hobson, E. W., 196
Hoheisel, G., 266
Hohenemser, K., 122
Horn, J., 266
Hort, W.-Thoma, A., 266

Ince, E. L., 266

Jacobi, C. G., 19
Jahnke, E.-Emde, F., 29, 154, 268

Kamke, E., 266
Keil, K. A., 72
Klein, F.-Haupt, O., 236
Kneschke, A., 266
Knopp, K., 14, 242, 268
Krall, G.-Einaudi, R., 122
Krylov, N.-Bogoliubov, N. N., 268

Lagrange, R., 196, 266
Langer, R. E., 142
Lappo-Danilevsky, A. I., 266
Lefschetz, S., 72, 266
Lense, J., 196
Levinson, N. (and Smith, O. K.), 77
Levy, H.-Baggott, A. E., 266
Liebmann, H., 266
Liénard, A., 76
Liouville, J., 10

Liouville, J.-Stekloff, V. A., 142, 145, 146, 147
Lonn, E. R., 53, 58, 71, 72

McLachlan, N. W., 266
Magnus, W.-Oberhettinger, F., 153, 179, 268
Malkin, I. G., 268
Milloux, H., 100
Minorsky, N., 77, 268
Mitropolski, I. A., 268
Murray, F. J.-Miller, K. S., 266
Myschkis, A. D., 266

Nemickii, V. V.-Stepanov, V. V., 266

Painlevé, P., 265
Peano, G., 10, 12
Perron, O., 58, 66, 71
Persico, E., 183
Petrovitch, M., 266, 267
Petrovski, I. G., 267
 -Landis, E. M., 76
Picard, E., 10, 209, 219, 236, 268
Picone, M., 136, 184, 268
 -Miranda, C., 98
Poincaré, H., 39, 53, 155, 241, 242
Pólya, G., 98
Poole, E. G. C., 267
Prodi, G., 149
Prüfer, H., 107

Reddick, H. W., 267
Richard, U., 100
Ritt, J. F., 267
Runge, C., 265

Sakarnikov, N. A., 66
Sansone, G.. 12, 72, 82, 149, 154, 155, 162, 183, 184, 186, 236, 267, 268
 -Conti, R., 87, 267
Sauter, F., 267
Schlesinger, L., 267
Serret, J. A. (and Scheffers, G.), 268
Slater, L. J., 260
Sonin, N. (and Pólya, G.), 98
Stepanov, V. V., 267
Sternberg, W., 255
Stoker, J. J., 268
Stolz, O., 156
Szasz, O., 265
Szegö, G., 98, 142, 183, 186, 268

Titchmarsh, E. C., 125, 267
Tricomi, F., 2, 6, 26, 28, 55, 75, 84, 87, 90, 91, 120, 125, 135, 136, 146, 147, 149, 153, 156, 157, 178, 186, 188, 189, 190, 201, 203, 204, 213, 219, 231, 260, 268
Trjizinski, W. J., 255, 266

de la Vallée Poussin, C., 92, 268
van der Pol, B., 76, 77, 92
Vessiol, 265
Vitali, G. (and Sansone, G.). See Sansone, G.

Watson, G. N., 154
Whittaker, E. T.-Watson, G. N., 236, 260, 268
Willers, A., 265
Wiman, A., 100
Wintner, A., 92

Zubov, V. I., 267

General Index

NOTATION: EL_2 denotes a linear second-order differential equation: the numbers refer to pages

addition theorem for circular functions, 17
Airy functions, 178
amplitude of elliptic functions, 22; sine-, cosine-, delta-amplitude, 22
asymptotic behaviour, 141, 202; in the complex field, 253; of eigenfunctions and eigenvalues, 166–180; of integrals of an EL_2 with no term in y', 148–162
asymptotic integration, 249 sqq; by series, for equations of Fuchs' type, 221
asymptotic representation of Bessel functions, 152–4, 262–4; confluent hypergeometric functions, 262; eigenvalues and eigenfunctions 129–132, 169–174; integrals of an EL_2, 152, 158, 166 sqq; Laguerre polynomials, 187–190; Legendre polynomials, 196–201; maxima and minima of integrals of an EL_2, 100
asymptotic series, 241 sqq

boundary conditions, 90, 109, 137
bounds for eigenvalues, 128, 129
brachistochrone, 89

canonical forms for a linear differential system, 40, 42, 44; for an EL_2, 165, 166
Carathéodory, solution of a differential system in the sense of, 6
cell (or hyperinterval), 5
centre, 36, 42, 44, 55, 64, 66, 71
characteristics of a first-order equation, 27 sqq; continuation of, 45–8; semi-characteristic, 46
circular functions, 13–9
col, 33, 41, 44, 48, 55, 66–71
cycle, 54; limit cycle, 66, 74–82

dependence of integrals on parameters, 11; initial values, 11
diagram of errors, 201, 264

differential equation: of Bessel, 106, 107, 152–4, 187, 228; of Clairaut, 2; of Laguerre, 181–3, 229; of Legendre, 191–4, 229; of Liénard, 76–82; of Papperitz, 234; of Riccati, 108; of van der Pol, 76; of the harmonic oscillator, 95, 126, 152: abridged, 39–45; homogeneous, inhomogeneous, 89; limit- 140; majorizing, minorizing, 114; satisfied by the hypergeometric function, 236; satisfied by the confluent hypergeometric function, 181, 258; self-adjoint, 96; totally Fuchsian, 228
discrete set, 95
distances between zeros of integrals, 82, 92, 104, 107

eigenfunctions, 117 sqq; normalized eigenfunctions, 175
eigenvalues, 117 sqq; negative, 119; for an equation with constant coefficients, 121, 126–7
elliptic functions, 19–26
elliptic integrals, 25, 26
equation: 'approximate', 142, 152, 166, 178; characteristic, 40, 140, 223, 243, 254; fundamental, 216; indicial, 223; integro-differential, 145. See also *differential equation, integral equation*
equivalence of a single equation and a differential system, 2, 31; a Sturm-Liouville system and an integral equation, 132–7
escape time, 46

first boundary-value problem, 109, 129
focus, 35, 42, 44, 48, 55, 58, 63–6
formula of Kummer, 138; Liouville, 157, 219
frequency of a vibrating string, 121
Fuchs' conditions, 219

functional transformation, 245
function-limit, 248
functions: Airy, 178; Bessel, 107, 153–4, 179, 187, 188, 258–9, 262–4; Green's function, 132–7; Riemann P-, 234 sqq; circular functions, 13–9; elliptic functions, 19–26; Gamma, 153, 188; hypergeometric, 237 sqq; confluent hypergeometric, 181, 237, 257 sqq; generalized hypergeometric, 240; of bounded variation, 162, 168; majorizing, 203; multivalent, 236

general integral, 1
Green's function, 132–7; symmetry of, 136, 137

harmonics, 122; harmonic modes of vibration, 121, 122
hyperbolic functions, 26, 95, 96

index of Poincaré, 53–5; index of a singular point, 54, 55
integral equations, 6, 132 sqq; of Fredholm type, 132 sqq; of Volterra type, 6, 132, 143 sqq

kernel of an integral equation. See *nucleus*
Kummer's formula, 138

lemma on integral equations of Volterra type, 149; the linear functional transformation T, 245–8, 250–2
limit cycle, 66, 74–82
limit equation, 140, 162
Liouville's formula, 157, 219
Lipschitz condition, 4

many-valuedness of integrals of an EL_2, 214–8
maxima and minima of integrals of an EL_2, 98–100
method: of successive approximations, 6–10, 146; Cauchy's method of majorizing functions, 202 sqq; Lagrange's method of variation of parameters, 89, 142; of Frobenius, 221 sqq; of Fubini, 142 sqq, 177; of Liouville-Stekloff, 142, 145 sqq
modulus of elliptic functions, 22; complementary modulus, 22
monotonic property of eigenvalues, 125, 126

node, 32, 34, 41, 44, 45, 48, 55–63; starshaped node, 33, 44, 53, 62
normalizing factor, 125, 175

nucleus of an integral equation, 136, 146; iterated, 147; resolvent, 147; symmetric, 136

O, o (Landau notation), 54
orthogonal system of functions, 124
orthogonality relations for eigenfunctions, 124; Laguerre polynomials, 184; Legendre polynomials, 195
oscillatory integrals, 96, 105

periodic solutions in the phase space, 82–8
periodicity conditions, 84, 90
periodicity of circular functions, 16; elliptic functions, 24
physical interpretation of eigenvalues, 120–3
points: conjugate points, 95, 105; singular points of a differential equation, 27 sqq; stationary points, 28; transition points, 177 sqq; of accumulation of eigenvalues, 119; of inflection, 28; Fuchsian singular points, points of regular singularity, points of non-essential singularity, 219
polynomials of Hermite, 183; Laguerre, 183–191; Legendre, 191–201
polynomials, zeros of orthogonal, 185
privileged directions, 33, 45
Prüfer change of variable, 107 sqq

recurrence formulæ for Laguerre polynomials, 186; Legendre polynomials, 196
reduced forms of an EL_2, 96–7, 163–6
reduction of differential equations to integral equations, 6, 132–7
regular point at infinity, 232
regular singular point, 219
regular singular point at infinity, 229
relaxation oscillations, 74–82

secular terms, 82
semi-characteristic, 46
separation of variables, 1, 120
series: asymptotic, 241; hypergeometric, 237; confluent hypergeometric, 259, 260; of inverses of eigenvalues, 129
simplicity of eigenvalues, 119; the zeros of integrals of an EL_2, 92
singular integral, 2
singularity: essential, 218; non-essential, 219
spectrum (*continuous*) *of eigenvalues*, 180, 181
stable integral, 148; focus, 66; node, 62, 63
sum of the roots of indicial equations of an EL_2, 231
supplementary conditions, 89

symmetric nucleus, 136
symmetry of Green's function, 136, 137
system: autonomous, 45; normal system of differential equations, 3, 202; Sturm-Liouville systems, 108 sqq; orthogonal systems of functions, 124

theorem: of Ascoli, 149, 151, 152; of Bendixson, on the nature of singular points, 48–53, on the existence of a closed characteristic, 75; comparison theorem of Sturm, 101–2; of numerical comparison, 103–4, for the function $\theta(x)$, 110–1; of existence and uniqueness, for differential systems, 5 sqq; of existence of eigenvalues, 117–9; of Fuchs', 219 sqq; of oscillation, 112–7; of separation of zeros, 102; of Sonin-Pólya, 98, 99; of de la Vallée Poussin, 92–5; on the zeros of a function given asymptotically, 170; on the existence of closed characteristics, 75; on the existence of periodic characteristics, 85; on the limit of y/y' in an EL_2, 155

topological correspondence, 54
transformation formulæ for hypergeometric functions, 239; confluent hypergeometric functions, 260
transformation, linear functional, 245
transition point of a differential equation, 177

variation of eigenvalues of an equation with variable coefficients, 125, 126
vibrating string, 120

zeros of circular functions, 16; elliptic functions, 25; integrals of an EL_2, 92, 98 sqq; orthogonal polynomials, 185; Laguerre polynomials, 184; Legendre polynomials, 196